T0138255

# Fuckology

# Fuckology

## CRITICAL ESSAYS ON JOHN MONEY'S DIAGNOSTIC CONCEPTS

Lisa Downing,
Iain Morland, and
Nikki Sullivan

The University of Chicago Press CHICAGO & LONDON

LISA DOWNING is professor of French discourses of sexuality at the University of Birmingham, United Kingdom. IAIN MORLAND works in music technology as an audio editor, sound designer, and programmer. NIKKI SULLIVAN is an honorary researcher in the Department of Media, Music, Communication, and Cultural Studies and teaches in the School of Communication, International Studies, and Languages at the University of South Australia.

The University of Chicago Press, Chicago 60637
The University of Chicago Press, Ltd., London

Printed in the United States of America

24 23 22 21 20 19 18 17 16 15 1 2 3 4 5

ISBN-13: 978-0-226-18658-0 (cloth)
ISBN-13: 978-0-226-18661-0 (paper)
ISBN-13: 978-0-226-18675-7 (e-book)
DOI: 10.7208/chicago/9780226186757.001.0001

Library of Congress Cataloging-in-Publication Data

Fuckology : critical essays on John Money's diagnostic concepts / Lisa Downing, Iain Morland, and Nikki Sullivan. — 1 Edition.
pages cm
Includes index.
ISBN 978-0-226-18658-0 (cloth : alkaline paper) — ISBN 978-0-226-18661-0 (paperback : alkaline paper) — ISBN 978-0-226-18675-7 (e-book) 1. Money, John, 1921–2006. 2. Sexology. 3. Gender identity. I. Downing, Lisa, author. II. Morland, Iain, 1978– author. III. Sullivan, Nikki, 1962– author.
HQ60.F83 2015
306.7—dc23
2014010241

♾ This paper meets the requirements of ANSI/NISO Z39.48-1992 (Permanence of Paper).

# CONTENTS

# ACKNOWLEDGMENTS

Lisa thanks the Leverhulme Trust for the award of a 2009 Philip Leverhulme Prize, which provided two years of research leave from the University of Exeter, and the Wellcome Trust for a Research Expenses grant that funded a trip to the Kinsey Institute, University of Indiana, in 2011 to carry out archival research in the John Money Collection. Thanks are due to the staff of the Kinsey Institute, especially Shawn C. Wilson and Liana Zhou, for their help in the course of this visit. Lisa is individually indebted to the following colleagues for providing—variously—information, ideas, feedback, references, and platforms for the dissemination of this research: Peter Cryle, Robby Davidson, Tim Dean, Robbie Duschinsky, John Forrester, Gert Hekma, Jennifer Burns Levin, Charles Moser, Dany Nobus, Eliza Steinbock, and Elizabeth Stephens.

For invaluable discussions and information, Iain thanks Neil Badmington, Diane Black, Jake Buckley, Helen D'Artillac-Brill, Milton Diamond, Alice Dreger, Katie Gramich, Laura Gregory, Katrina Karkazis, Emma Ralph, Keith Sigmundson, and Richard Symes. For feedback on earlier versions of chapter 3, Iain thanks Robert Eaglestone, Mandy Merck, Margrit Shildrick, and Joanna Zylinska. Chapter 3 originates in research that was supported by the UK Arts and Humanities Research Council, and includes material previously published as "Plastic Man: Intersex, Humanism and the Reimer Case," *Subject Matters* 3, no. 2/4, no. 1 (2007): 81–98, which is reproduced here with permission of London Metropolitan University, and with thanks to Paul Cobley.

Nikki is grateful to all those who have helped her to think through the material covered in chapters 1 and 5, in particular, Kellie Greene, Susan Stryker,

Samantha Murray, Jessica Cadwallader, Elizabeth Stephens, Dennis Bruining, Rosalyn Diprose, Sara Ahmed, and, of course, Iain and Lisa. She would also like to thank the Special Collections staff at the University of Otago Library, Dunedin, for help navigating the collection of John Money's manuscripts and papers.

All three authors would like to thank the team at the University of Chicago Press, especially Douglas Mitchell and Tim McGovern, and the expert readers, Ivan Crozier and Susan Stryker, for their constructive feedback on this manuscript.

INTRODUCTION

# On the "Duke of Dysfunction"

*Lisa Downing, Iain Morland, and Nikki Sullivan*

> Dr. John Money is the Duke of Dysfunction, a man who writes about "unspeakable" human sexual problems with such dignity and care that his case histories make me feel almost normal.
>
> JOHN WATERS, jacket endorsement for John Money, *Gendermaps: Social Constructionism, Feminism, and Sexosophical History*

The New Zealand-born, US-based psychologist John Money (1921–2006) has had a singular influence on the diagnosis and treatment of (to use Money's terms) "hermaphroditism," "transsexualism," and "paraphilia."[1] The reception of his more than five hundred articles and over forty books, as well as hundreds of neologisms including "gender" itself, has been both exceptionally significant and strikingly uneven.[2] Whereas "gender" is now a ubiquitous, everyday term in the English-speaking world, and "lovemap" has entered the lexicon of popular psychology, some of his more outlandish coinages, concepts, and recommendations have entered neither popular nor medical currency.

Money's widespread yet disparate uptake is explained partly by the fact that his stylistically bizarre texts were aimed at multiple audiences, most often physicians, psychiatrists, and sexologists, but sometimes anthropologists, historians, psychoanalysts, and lay readers. Money's career was also beset by ethical controversy, exemplified by the internationally publicized case of David Reimer. Following sex reassignment in infancy under Money's guidance, in response to a circumcision accident, Reimer's story was held variously to show Money as humane and barbaric, naive and deceitful, a social constructionist and an anatomical determinist. Just as Money's ideas have been characterized as either pathologizing or liberating, so too has Money's flamboyant persona been beatified or damned. These tendencies to polarize Money and his work are both productive and symptomatic of a failure to interrogate the complexities, contradictions, and tensions in Money's œuvre. Therefore, a careful cross-disciplinary, multiauthored engagement with Money's work and its deployment is overdue.

For example, sufficiently close attention has not hitherto been paid to Money's fears (which were probably understandable given the historico-political context of his work during the 1950s) that sexology could be dismissed as a prurient, if not altogether "perverse" practice, much as its forerunners in the nineteenth century had been demonized by many doctors and clerics. It is in this light that we understand Money's constant demand that his field, for which his predecessors "could not find a name," should be considered both a legitimate science and the natural home of taboo-busting "sex research."[3] The conception of a unique (nameless) scientific field—which Money argues complements other sciences such as urology, gynecology, endocrinology, and so on, and for which he suggested the name "fuckology"—functions hand in hand with Money's fascination with a "Linnaean" taxonomic approach to human experience.[4] A passion for creating taxonomies is evident in his coining of a plethora of diagnostic and technical terms, including "Adam Principle," "exigency theory," "gynemimesis," "mindbrain," "neurocognitional," "normophilia," "phylism," "troopbondance," and a whole range of paraphilias, such as "apotemnophilia," "autassassinophilia," and "autonepiophilia," to give only a sample of those at the beginning of the alphabet. It is, then, somewhat ironic that Money claimed inspiration from Willa Cather and Ernest Hemingway for their "economy of words and uncluttered style."[5]

Connecting Money's aspiration to scientificity and his tendency to taxonomic invention is his view of both "gender" and the "lovemap" as kinds of "native language."[6] In one paper from 1982, Money wrote: "For sexological research the development of . . . an analytic vocabulary is not simply an ideal, but an absolute necessity, for without it erotosexual practice cannot be properly subdivided and reduced to identifiable units for investigation in research."[7] And, in the same paper, he could not resist adding, after a mention of "every behavioral unit," a parenthetical nascent term for this concept: "(behavioron)."[8] For Money then, as we will explore in this book, the acquisition of a language about sex was an object of study, a scientific method, and a master metaphor, all at once.

The title of this book, *Fuckology*, is a reappropriation of a neologism that Money proposed for introduction into scholarly, clinical, and lay discourse. He wrote in 1988 of the need for "a word like fuckology, used in everyday, vernacular English to signify the science of what it is that people actually do under the cover of polite expressions like making love or having sex."[9] Although this book's critical remit is wider than Money's contribution to the study of sexual orientations and practices, and while the book certainly does not seek to further Money's agenda by using "fuckology" as a candid descriptor in the way that he suggests is possible, the term "fuckology," used against the grain, strikes us as extraordinarily appropriate shorthand to describe a method of queering—or fucking with—sexology, and with the logic of scientificity in which it is invested. In particular, we are aware of, and seek precisely to exploit, the readerly discomfort and uncertainty potentially engendered by the use of this nonacademic vocabulary (the sort of vocabulary that Money himself might have called "the terminology of the barnyard").[10] Money often used sexualized rather than clinical terms for sexual activities, such as the verbs "to quim" and "to swive," which he derived from vernacular seventeenth-century terms for genitalia, and which were intended by Money to describe "the active-assertive practice of the female and the male, respectively, in penovaginal copulation."[11] He seems to have understood such linguistic misdemeanors as acts of daring resistance to an imagined censorious "sex police," albeit in a gesture that risked undermining his claims made elsewhere for the scientific seriousness of this work; Money was aware as early as 1955 that neologisms could be regarded as "perverse technical jargon."[12] Putting aside our various reservations about Money's intentions in introducing neologisms such as these, we find the term "fuckology" productive insofar as it suggests resistance to a unified theory of Money. Further, "fuckology" disrupts the domestication of Money's peculiar œuvre as a transient moment along a path to ever more scientific and humane knowledge of "hermaphroditism," "transsexualism," and "paraphilia."

## BEING DR. MONEY

As alluded to above, much of the available commentary on John Money casts him as either a god or a monster. While this book eschews a psychologizing "man and his works" approach to Money's contribution, in favor of an interrogation of his influences and contexts in producing and transforming the diagnostic concepts with which he worked, it will be necessary to examine the associations that accrue to the name "John Money" in the course of his career, and in the posthumous reception of his work. Moreover, while our book seeks to reach conclusions about Money's contributions to sexology without adhering to the personal loyalties and enmities that led to his deification/demonization, it would be erroneous to ignore the ethically and politically charged circumstances and environment that produced such work. To this end, we offer the following short biographical sketch.

John Money was born on July 8, 1921, in Morrinsville, New Zealand, to an Australian father and an English mother. He completed high school early and went on to study at Victoria University, Wellington. In 1944, he graduated with a teacher's certificate and a double master's degree in education and philosophy/psychology. He took up an appointment as a junior lecturer in the psychology department at the University of Otago in Dunedin, where he worked for three years. Because, at that time, it was not possible to read for a doctorate in psychology in New Zealand, Money emigrated to the United States in 1947, working as a psychology resident in a Pittsburgh hospital before being accepted into a PhD program at Harvard University in the Department of Social Relations. He graduated in 1952 having produced a dissertation titled "Hermaphroditism: An Inquiry into the Nature of a Human Paradox."[13] Money did not train as a medical doctor, surgeon, or psychiatrist, as has sometimes been assumed.[14]

His clinical work with intersex individuals began before he had even completed his doctorate, initially at Massachusetts General Hospital and Harvard Medical School, and subsequently at the Johns Hopkins Harriet Lane Home for Invalid Children. Money moved to Johns Hopkins University in 1951 with his mentor, the psychologist and physician Joan Hampson. Shortly afterward, he met Lawson Wilkins, MD, influential head of the Clinic for Pediatric Endocrinology at Johns Hopkins. Wilkins allowed Money to interview patients at the clinic for his doctoral research, and once the research was completed, Money continued to work under Wilkins, in close collaboration with Joan Hampson and her husband John (also a psychologist and physician).[15] Together they composed what Money would later call the "Psychohormonal Research Unit," affiliated with both the Departments of Pediatrics and Psychiatry.[16] The import of the work by Money and the Hampsons in shaping protocols for intersex treatment cannot be overstated. Yet, despite their pioneering collaboration, Money and John Hampson "were no longer on speaking terms" by 1957.[17] The definitive reason for this rift is not given in any published literature on Money that we have been able to find.

By the mid-1960s, Money's interest had turned to transsexualism and the possibility of surgical treatment: between 1964 and 1967 he was part of a research team led by Harry Benjamin (and including Ruth Rae Doorbar, Richard Green, Henry Guze, Herbert Kupperman, Wardell Pomeroy, and Leo Wollman), whose study of transsexualism was funded by the Erickson Educational

Foundation. The latter had been established in 1964 by the wealthy trans man and patient of Benjamin's, Reed Erickson. Research undertaken by the group was integral to the official establishment of the Johns Hopkins Gender Identity Clinic (in July 1966) as well as to the formation of the Harry Benjamin Foundation (in 1967). Moreover, according to Benjamin, Money "was probably more responsible than any other individual for the decision that such an august institution as Johns Hopkins Hospital would . . . endorse sex-altering surgery in suitable subjects," a practice for which, at the time, there was little support among medical professionals.[18] Money also served on the advisory board of the Harry Benjamin Foundation, which regularly referred patients to Hopkins, ensuring a client base for the treatments he was developing.[19] However, Money later expressed disappointment that the Gender Identity Clinic, whose name he professed to have inspired, did not "become a center for manifold syndromes related to gender identity," and remained focused on transsexualism until its closure in 1979.[20]

In the 1980s and 1990s, John Money wrote widely about the paraphilia diagnosis and his advocacy for the use of antiandrogen medication in the treatment of both sex offenders and, controversially, other "paraphiles" with non-offending behavior. In a 1987 paper, Money claimed that he had been studying the use of the drug Depo-Provera (medroxyprogesterone acetate) with sex offenders at Johns Hopkins since as early as 1966,[21] at which time the drug had not been approved for that usage.[22] The uptake of his combination of drug therapy and "talking therapy" throughout the United States and Europe was intermittent, but not insignificant.[23] Additionally, Money's interventions in debates about pedophilia, arguing that there is a clinical distinction to be drawn between "affectional pedophilia" and "sadistic pedophilia," and appearing ambivalently supportive of elements of the propedophilia movement,[24] led controversy to dog his reputation, a taint on his name that would become indelible once the outcome of the Reimer case was a matter of public knowledge. David Reimer (pseudonymously at first) spoke publicly in 1997 about his surgery and sexological treatment with Money; John Colapinto published his critical book on the case in 2000; and Reimer ended his own life in 2004. Money remained at Johns Hopkins for the duration of his career, supported by numerous grants from organizations including the Josiah Macy Jr. Foundation and the National Institute of Child Health and Human Development.[25] In his final years, a somewhat discredited Money suffered from progressive dementia.[26] He died in 2006, of causes related to his Parkinson's disease, one day before his eighty-fifth birthday.[27]

The bare facts of his life and work aside, what can be learned of John

Money's biography is partial and inevitably biased. The range of sources that, for different purposes, describe his personality and career, are heavily colored by the tastes and political affiliations of the given author. For example, few biographical sources make mention of Money's marriage in the 1950s unless to lead us to infer from its failure something about his character. Thus, in his critical journalistic account of the Reimer case, Colapinto writes: "As an adult, Money would forever avoid the role of 'man of the household.' After one brief marriage ended in divorce in the early 1950s, he never remarried and has never had children."[28] It would be easier to feel indignation on Money's behalf for this implicit accusation that he was not a "mature" "responsible" man, based on a normative notion of the functional, (re)productive citizen, were Money himself not responsible for producing similar charges about patients, especially paraphiliacs, as will be explored in some of the chapters that follow.

Indeed, it is striking that the accusations of "perversion" leveled at Money by his detractors, for his ambivalence to pedophilia and his own unconventional sexual behaviors, are not dissimilar to some of Money's more scathing observations regarding what he deemed inappropriate sexual behaviors (while writing approvingly of those activities he is known to have enjoyed, including nudism and group sex, practiced in the libertarian environment of the Society for the Scientific Study of Sexuality's [SSSS] gatherings).[29] Thus it is unhelpful to see Money entirely as either a misunderstood defender of sexuality in all its forms or as the victim of deliberate misinformation. He advocated both "the study of human sexuality in its recreational as compared with its purely procreational function," in which the SSSS led the way, Money claimed,[30] and the teaching of children that "sex differences are primarily defined by the reproductive capacity of the sex organs."[31] Money never synthesized such views. Consequently, the uneven reception of his work reflects a certain truth about its character as an "ideological octopus" (as one of us has written elsewhere), appealing in different ways to conservative and progressive commentators alike.[32] We argue that Money's work is not so much "interactional," contrary to what he frequently asserted, but tentacular.[33]

Accordingly, angry responses to Money in the wake of Colapinto's revelations about the Reimer case issued from both those apparently vindicated champions of biological explanations of gender, who perceived Money's constructionist experiment to do violence to the unassailable nature of "man" and "woman," and from liberal social constructionists alarmed at the ethics of Money's practice. So, in a letter to the *Washington Times*, Carey Roberts (described as a "writer and media analyst") wrote that Colapinto's biography of Reimer "revealed the psychologist to be a charlatan, tireless self-promoter and intellectual fraud." He added that "the feminist dogma that gender is socially constructed remains widespread in our society. Boys receive constant messages they should act more like girls. David Reimer's sad story should cause us to reconsider our mass experiment in gender re-education."[34] In politically contrasting terms, but revealing equal vilification for Money, Mark Cochrane wrote in the *Vancouver Sun* that "neo-Darwinist explanations for gender identity reinforce a rigidity of roles and expectations, and they can be used to justify anything, from rape to an array of social and domestic inequities. . . . For these reasons, and despite *the existence of monsters such as Dr. Money*, the 'nature' side of the debate continues to produce a more dangerous rhetoric. Unfortunately, Colapinto's book may play into that."[35] The proconstructionist writer refers to Money using an othering, teratological label in order to exculpate the "nurture" "side" of the debate from being tainted by association. And, to neatly complete this discursive circle, following the announcement of Reimer's death, David Jones would write in the British *Daily Mail*, espousing the view of gender that one would most readily associate with that newspaper, "Arrogant to the last, Dr Money still refuses to acknowledge that the cruel human experiment he devised to confirm his flawed ideology has been a monstrous failure."[36] While disagreeing with Cochrane about the value of the "ideology" in question, he repeats the familiar vocabulary of monstrosity to apply, albeit indirectly in this syntactic construction, to John Money.

To Money's defenders and friends, however, all such criticisms are liable to be interpreted as misrepresentations of Money's noble project, or attempts to scapegoat a brave pioneer whose ideas could be "simply too intellectually demanding to pursue."[37] Anke Ehrhardt, whose doctoral research took place at Money's Psychohormonal Research Unit,[38] writes in her obituary of Money that "it was . . . regrettable that [over the last decade] John Money's work was often globally criticized and rejected and that he as a person was unjustly scapegoated." She goes on:

> When we talked about the attacks [on his reputation, etc.], I tried to reassure him that he would share the fate of many truly pioneering giants in science, namely, that we were experiencing a swinging of a pendulum that ultimately would swing back and that his work would find the proper place in history. Indeed, the pendulum has already started to swing back to give John Money the proper credit for his extraordinary contribution to the field of psychoendocrinology and sex research.[39]

And Richard Green writes in his obituary of Money that "John's last years were doubly tragic." In addition to Money's progressive dementia, "detractors had it appear that he set about amputating a boy's penis so he could test his theory. He was denounced as a Dr. Mengele on Australian TV's '60 Minutes.' Newspapers that should know better, such as the *New York Times*, failed to provide balanced reporting."[40] While the "Mengele" analogy is obviously exaggerated and emotive, it is unfortunate that Green should focus on Money's ruined reputation as the most regrettable aspect of what happened to Reimer.

In a celebratory piece written to mark Money's seventieth birthday, "For the Sake of Money," Paul R. Abramson describes Money as "arguably the most prominent (and prolific) sex researcher of our day." He goes on: "For nearly forty years, John has produced an extraordinarily impressive scholarly record, on topics ranging from gender identity and gender role to sexual orientation and the freedom of sexual expression. Perhaps even more important, however, have been his theoretical contributions, which utilize an interdisciplinary perspective in conceptualizing the development and expression of human sexual behavior."[41] Abramson's claim that interdisciplinarity is central to Money's method is an important one, and one deserving of interrogation. Money was obviously fascinated by, and to some extent versed in, numerous and diverse scholarly and medical fields. However, his style can appear to disavow the very character of interdisciplinary work: the potential for interdisciplinarity to relativize the truth content of a given disciplinary stance by offering insights from another body of thought or method is never fulfilled in Money's work. He is a relentlessly dogmatic writer. Although, as we shall explore, Money occupies starkly different positions with regard to, for example, the import of nature and nurture, neural and social factors, at different stages in his career and sometimes within individual publications, Money seeks at every point in his œuvre to argue for the rigor and definitional nature of his statements as they stand at that moment. This is especially apparent in the following claims about the influence of hormones on gender and sexuality.

In 1961, Money asserted that "the sex hormones, it appears, have no direct effect on the direction or content of erotic inclination [which, for Money, was an aspect of gender] in the human species. These are assumed to be experientially determined."[42] Four years afterward, the biologist Milton Diamond published a critical review of work by Money and the Hampsons, which included a counterclaim that "when we consider prenatal as well as postnatal existence, hormones may be regarded as directional as well as activational; and at birth the individual may be considered to have been neurally predisposed by genetic and hormonal means toward [identification as] one sex."[43] In 1971, Money called Diamond's paper "a rather ill-considered critique" by "an inexperienced student of biology."[44] Yet, two years later, in a paper that did not cite Diamond, Money claimed that research from as far back as the late 1950s showed that "fetal gonadal hormones . . . have an influence on neural pathways in the brain." He added, "If I had said that even as recently as 10 years ago, people would be wanting to put me away. To imagine that fetal gonadal hormones could have anything to do with brain pathways!"[45] Notwithstanding this apparent convergence with Diamond's position, Money continued in subsequent decades to refer to the former's 1965 paper as "a lengthy polemic."[46] To admit of the potential of being wrong, or to settle for the productive tension of ambiguity, is not a feature of Money's rhetorical range, even if, as the essays in this book will show, Money's claims are often brazenly inconsistent. If Abramson is right, though, that Money is an eminently interdisciplinary thinker, it makes the project of this book all the more urgent and justifies its methodological sweep, because we approach Money's texts from outside the disciplines in which he wrote, and from a range of perspectives, in order precisely to evaluate them from the multiple and intersecting viewpoints of history, critical and ethical theory, gender/sexuality studies, and textual analysis.

## THE PURPOSE OF THIS BOOK

This book constitutes the first theoretical humanities-based analysis of John Money's contribution to the three diagnostic concepts "hermaphroditism," "transsexualism," and "paraphilia." It interrogates continuities and discontinuities between the concepts, to illuminate the causes, consequences, and power dynamics of Money's work. Following this coauthored introduction, the book falls into two parts, with names inspired by Money's book *Vandalized Lovemaps*. The first of these, "Mapping," is primarily historical and contextual, charting Money's sometimes disavowed debt to earlier sexological and psychological models, and situating his work in its contemporary moment. The second part, "Vandalizing," takes Money's ideas into new critical domains, placing his texts in dialogue with a range of relevant theories and discourses. There are three single-authored chapters in each section, each of which focuses principally but not exclusively on one of the three diagnostic concepts—"transsexualism" in chapters 1 and 5, written by Nikki Sullivan; "paraphilia" in chapters 2 and 6, written by Lisa Downing; and "hermaphroditism" in chapters 3 and 4, written by Iain Morland. Downing's contribution to the book issues primarily from a post-Foucauldian, history-of-medicine perspective; it is also informed by her specialism in continental philosophy and critical theory, incorporating insights from ethical theory, poststructural psychoanalysis, feminist theory, and queer theory. Morland's contribution is characterized by a multidisciplinary engagement with science and technology studies, continental philosophy, and gender studies, and a specialism in the ethics, psychology, and politics of intersex. Sullivan's contribution is located primarily at the intersection of queer theory, body modification studies, and feminist philosophy of the body, and presents a critical exploration of the technologized character of bodily-being-in-the-world. So, rather than proceeding from the discipline of sociology of science (for the social determinants of Money's knowledge claims are not our main concern), or feminist science studies in the empiricist sense (for we are not trying to "do science" at all), the chapters combine close reading and contextualization of Money's corpus with sophisticated and eclectic theorization.

The first part of the book opens with Sullivan's chapter, "The Matter of Gender," on Money's establishment of the model of gender identity/role (G-I/R), which has been at once highly influential and largely misunderstood. Given that his conception of G-I/R was foundational to his understanding and "treatment" of intersex, transsexuality, and what he saw as "atypical" erotic desires and practices, as well as to his critique of feminism(s), liberationism, and other critical movements, it is clear that a close reading of Money's account of G-I/R and its uptake (by, for example, radical feminists, liberal feminists, gay liberationists, queer theorists, trans theorists/activists, feminist philosophers of the body, and new materialists) is long overdue. This chapter provides a close critical engagement with Money's often contradictory claims about G-I/R and its attribution, arguing that while neither simply constructionist nor purely deterministic, Money's formulation of G-I/R exemplifies what the feminist scholar Linda Nicholson has referred to as "biological foundationalism."[47]

In a similar vein, chapter 2 by Downing, "A Disavowed Inheritance: Nineteenth-Century Perversion Theory and John Money's 'Paraphilia,'" traces a genealogy from foundational nineteenth-century sexological "perversion" theory, wherein sexual deviance is posited as a dangerous symptom of a degenerate society, through twentieth-century psychoanalytic writing on "perversion," to John Money's authoritative writings on "paraphilia" in the 1980s and 1990s. It argues that although Money's texts explicitly reject the values and systems of their historical forebears, the logic and rhetoric of nineteenth-century sexology haunt Money's thinking. Given the influence of Money's involvement in the definition of the paraphilia diagnosis (such as his naming of scores of paraphiliac conditions; his contribution to the *DSM-III-R* in 1987; his pioneering and controversial treatment program for paraphiliac sex offenders), the chapter contends that the disavowed inheritance of nineteenth-century sexological values in twentieth-century psychiatry continues to play into formulations of "paraphilia" today.

In chapter 3, "Gender, Genitals, and the Meaning of Being Human," Morland explains Money's unique impact on the medical treatment of intersex by situating his work in relation to twentieth-century discourses of humanism. Arguing that Money invoked humanism in his signature claims that gender and genitals are malleable in infancy, the chapter shows how Money borrowed authority from contemporary scientific theories of human adaptability. His purported humanism had the mutually reinforcing effects of facilitating the uptake of Money's ideas about intersex, while also instituting gender as a core human quality, flexible by definition. The chapter details the confluence in Money's work of humanist discourses from across the century: psychologist Alfred Adler's inferiority theory; the popularization of cosmetic surgery after World War I; the post–World War II scientific rejection of race; the medical recognition and treatment of transsexuality; and changing sexual norms during the 1960s. It explores how these discourses converged in Money's test case for intersex treatment, the story of David Reimer. Further, the chapter reveals the surprising persistence in critiques of Money of claims about the plasticity of human nature.

The second part of the book opens with chapter 4, "Cybernetic Sexology," in which Morland engages with Money's avowal to think cybernetically about sex, gender, and sexuality. Using a cybernetic vocabulary of variables, thresholds, and feedback systems, Money claimed to offer a more scientific and up-to-date sexology than hitherto possible. This chapter presents the first-ever critical analysis of Money as a cybernetic theorist. Evaluating the context and rhetoric of his claims, the chapter explains how Money used cybernetics—the study of communication and control, conceived during 1940s military research—to distance sexology from both psychoanalytic and biological studies. Further, it shows how cybernetic theory shaped Money's treatment recommendations for individuals with ambiguous genitalia: he drew both explicitly and indirectly on cybernetic thinkers such as mathematician Norbert Wiener and psychiatrist Ross Ashby. However, the chapter also critiques a formative error made by Money in his application of cybernetics to sexology. Cybernetics theorized dynamic systems that can adapt, not merely repeat. It was therefore irreconcilable with the sudden, irrevocable establishment of gender in infancy that was axiomatic for Money. Consequently, the chapter argues that Money's model of gender, as reiterated without conscious reflection, was closer to a psychoanalytic view of unconscious determinism than a cybernetic theory.

In chapter 5, "Reorienting Transsexualism: From Brain Organization Theory to Phenomenology," Sullivan examines the claim, central to Money's later work on G-I/R, that the brain is the site of the unification of "biological and social determinism," the place where schemas of gender identification and complementation become locked in, and from whence they function as "templates in the governance of sex dimorphic behavior."[48] In order to substantiate this view of dimorphic brain schemas, and more particularly the role of prenatal hormones in their development, Money turns to sociobiological accounts of animal sex. Drawing on the work of feminist science studies scholars such as Helen Longino, Anne Fausto-Sterling, Marianne Van Den Wijngaard, and Rebecca Jordan-Young, this chapter critically interrogates the role and function of brain organization theory, studies of "animal sex," and the relation between them in Money's articulation(s) of transsexualism. While it discusses the relatively small number of articles Money published on transsexualism, it focuses primarily on an article, coauthored with John G. Brennan, entitled "Heterosexual vs. Homosexual Attitudes: Male Partners' Perception of the Feminine Image of Male Transexuals" (1970), since this article explicitly brings these issues and approaches together in order to make a series of troubling heteronormative claims about trans women—claims that thoroughly undermine what some might see as Money's championing of trans bodies and practices. The chapter also offers a phenomenological account of bodily-being-in-the-world, and of (dis)orientation, as a counter to the limits affected by Money's turn to brain organization theory. The aim of this turn to feminist phenomenology is to indicate how the potentialities of Money's work on sex, gender, and sexuality might be articulated otherwise.

In chapter 6, "'Citizen-Paraphiliac': Normophilia and Biophilia in John Money's Sexology," Downing focuses in detail on one key aspect of Money's work on paraphilia that was introduced in chapter 2, namely, the idea that paraphilia is the dangerous and deadly form of eroticism that compromises the proper, life-giving "nature" of reproductive sexuality and thereby threatens the social order. It argues that, at the fantasy level of Money's system, paraphilia is understood as leading to individual and social death, in the intellectual historical context of a biological and psychological worldview in which the theory of "biophilia" was proliferating. The chapter explores Money's wish to attain a "paraphilia-free" society,[49] alongside his paradoxical call for "a pluralistic democracy of sexualism,"[50] showing that his writing on paraphilia reveals one of the tensions at the heart of his "sex-positive" liberalism and libertarianism. For Money, and many who influenced and follow him, the possibility of the figure of the "paraphiliac citizen" is not admitted. Using insights from queer citizenship studies, the antisocial turn in queer theory, and what Judith Halberstam has termed "shadow feminism,"[51] the chapter undertakes a critique of the bionormativity of this premise and the conservative paradigm of health and harm on which it rests.

*Fuckology* closes with a coauthored critical conclusion, "Off the Map," which identifies and examines some of the key features of Money's textuality and practice that interconnect his interventions into all three diagnostic concepts. These include his metaphor and methodology of "mapping," his insistence on the power of language acquisition, and his concomitant obsession with neologistic labeling, especially of the phenomenon known as gender. The conclusion critiques Money's fears that without unanimity in language, sexology could collapse into "wasteful word games,"[52] just as gender development might become "ambiguous" in the manner of "native bilingualism."[53] Yet, it shows too that Money's unremitting invention of terms stoked these very fears, demarcating a private linguistic world in which he and his colleagues were the only native speakers. So, if each of us "fucks with" Money in a slightly different way in this book, our contributions are brought together by their determination to do critical justice to the "fucked-up-ness" of Money's texts, contradictory, repetitive, and dysfunctionally self-undermining as they are.

## NOTES

1. "Hermaphroditism" is more often referred to today as "intersex" or "intersexuality," or in recent medical terminology—and controversially for some critics and activists—"disorders of sex development." "Transsexualism" is often referred to today as "transgender" or just "trans"/"trans*," umbrella terms suggesting a range of identities from genderqueer to post-operative trans men and women. "Paraphilia" is the contemporary psychiatric term for what was formerly (and still is, in psychoanalytic circles) referred to as "sexual perversion." Self-identifying "paraphiliacs" might talk of "kink" rather than paraphilia, and describe themselves as "kinksters" or "pervs."

2. Interviewed in 1996, Money claimed to have "published, as author or editor and contributor, thirty nine books and five hundred odd scientific journal articles and book chapters." Further books by Money appeared since then. In an obituary, Anke Ehrhardt stated that Money published "close to 2,000 scientific articles, books, chapters and reviews." See "Amazon.com Talks to John Money," 1996, archived at http://web.archive.org/web/19991218105428/http://www.amazon.com/exec/obidos/show-interview/m-j-oneyohn, para. 2 of 10; Anke A. Ehrhardt, "John Money, Ph.D.," *Journal of Sex Research* 44 (2007): 223–24, 223.

3. John Money, "Sexology: Behavioral, Cultural, Hormonal, Neurological, Genetic Etc.," *Journal of Sex Research* 9 (1973): 1–10, 3.

4. John Money, *The Psychologic Study of Man* (Springfield, IL: Charles C. Thomas, 1957), v.

5. "Amazon.com Talks to John Money," para. 4 of 10.

6. John Money, "Gender: History, Theory, and Usage of the Term in Sexology and Its Relationship to Nature/Nurture," *Journal of Sex and Marital Therapy* 11 (1985), 71–79, 76; John Money, *Lovemaps: Clinical Concepts of Sexual/Erotic Health and Pathology, Paraphilia, and Gender Transposition in Childhood, Adolescence, and Maturity* (Buffalo, NY: Prometheus, 1986), 7, 12, 17, 19, 118, 209.

7. John Money, "To Quim and to Swive: Linguistic and Coital Parity, Male and Female," *Journal of Sex Research* 18 (1982): 173–76, 175.

8. Money, "To Quim and to Swive," 175.

9. John Money, *Gay, Straight and In-Between: The Sexology of Erotic Orientation* (Oxford: Oxford University Press, 1988), 4–5.

10. Money, "To Quim and to Swive," 173.

11. Money, "To Quim and to Swive," 173.

12. John Money, "Linguistic Resources and Psychodynamic Theory," *British Journal of Medical Psychology* 28 (1955): 264–66, 264.

13. See John Money, *Venuses Penuses: Sexology, Sexosophy, and Exigency Theory* (Amherst, NY: Prometheus, 1986), 5, and Ehrhardt, "John Money, Ph.D.," 223.

14. For example, he was bylined as "John Money, M.D." in "Nativism versus Culturalism in Gender-Identity Differentiation," in *Sexuality and Psychoanalysis*, ed. Edward T. Adelson (New York: Brunner/Mazel, 1975), 48–64, 48.

15. See Money, *Venuses Penuses*, 8; John Money, *Sin, Science, and the Sex Police: Essays on Sexology and Sexosophy* (Amherst, NY: Prometheus, 1998), 310.

16. Ehrhardt, "John Money, Ph.D.," 223. The name "Psychohormonal Research Unit"—another of Money's neologisms—appears to have been applied retrospectively by Money to his earliest work in Wilkins's clinic; in a 1975 paper, he dated the unit's inception to 1951; however, none of his 1950s papers regarding research conducted at the clinic used the phrase. See John Money, "Human Behavior Cytogenetics: Review of Psychopathology in Three Syndromes—47,XXY; 47,XYY; and 45,X," *Journal of Sex Research* 11 (1975): 181–200, 181.

17. Richard Green, "John Money, Ph.D. (July 8, 1921–July 7, 2006): A Personal Obituary," *Archives of Sexual Behavior* 35 (2006): 629–32, 629.

18. Harry Benjamin, "Introduction," in *Transsexualism and Sex Reassignment*, ed. Richard Green and John Money (Baltimore: Johns Hopkins University Press, 1969), 1–10, 7.

19. Joanne Meyerowitz, *How Sex Changed: A History of Transsexuality in the United States* (Cambridge, MA: Harvard University Press, 2002), 219.

20. Money, *Sin, Science, and the Sex Police*, 121. The clinic closed following a critical (and contentious) review of its outcomes. See Jon K. Meyer and Donna J. Reter, "Sex Reassignment," *Archives of General Psychiatry* 36 (1979): 1010–15.

21. John Money, "Treatment Guidelines: Antiandrogen and Counseling of Paraphilic Sex Offenders," *Journal of Sex and Marital Therapy* 13 (1987): 219–23, 219. The drug had also

been prescribed by clinicians to prevent miscarriage, which according to Money could have the "rare, paradoxical, and unexplained" effect of creating "a masculinized clitoris" in a female fetus: in other words, in a little-known convergence between two of Money's diagnostic concepts, the same drug that he advocated for treating "paraphiles" caused intersex in infants. See John Money and Anke A. Ehrhardt, *Man and Woman, Boy and Girl: The Differentiation and Dimorphism of Gender Identity from Conception to Maturity* (Baltimore: Johns Hopkins University Press, 1972), 288.

22. See Daniel Tsang, "Policing Perversions: Depo-Provera and John Money's New Sexual Order," in *Sex, Cells, and Same-Sex Desire: The Biology of Sexual Preference*, ed. John P. De Cecco and David Allen Parker (Philadelphia: Haworth, 1995), 397–426, 402.

23. See Tsang, "Policing Perversions," 402, for an account of legislative decisions to authorize this kind of treatment in various states.

24. As Tsang states, "[Money] testified for the editors of the *Body Politic*, defending their publication of an essay on pedophilia. He wrote a favorable introduction to Theo Sandfort's *Boys on Their Contacts with Men*. . . . In addition, he has been sought out and interviewed favorably by pedophile movement publications such as *Paidika* and *OK*, both based in the Netherlands" ("Policing Perversions," 411). Relatedly, Money has claimed that "when a grandfather fondles his own beloved grandchild while sleeping in the same bed, the act is not incestuous in the same sense as when a visiting uncle forces his screaming, terrified, new pubertal niece to copulate with him" (John Money, "Paraphilias," in *Handbook of Sexology*, ed. John Money and Herman Musaph [Amsterdam: Elsevier/North Holland Biomedical, 1977], 917–28, 922).

25. Lawson Wilkins, *The Diagnosis and Treatment of Endocrine Disorders in Childhood and Adolescence*, 2nd ed. (Springfield, IL: Charles C. Thomas, 1957), xi; Ehrhardt, "John Money, Ph.D.," 223.

26. Green, "John Money, Ph.D.," 631.

27. Ehrhardt, "John Money, Ph.D.," 223.

28. John Colapinto, *As Nature Made Him: The Boy Who Was Raised as a Girl* (London: Quartet, 2000), 27.

29. See, for example, Green, "John Money, Ph.D.," 630; John Money, "Determinants of Human Gender Identity/Role," in *Handbook of Sexology*, ed. John Money and Herman Musaph (Amsterdam: Elsevier/North Holland Biomedical, 1977), 57–79, 76.

30. John Money, "The Development of Sexology as a Discipline," *Journal of Sex Research* 12 (1976): 83–87, 83.

31. Money and Ehrhardt, *Man and Woman*, 14.

32. Iain Morland, "Editorial: Lessons from the Octopus," *GLQ* 15 (2009): 191–97, 195; the cephalopodan metaphor is borrowed from Justin Lewis's *The Ideological Octopus: An Exploration of Television and Its Audience* (New York: Routledge, 1991).

33. John Money, "Critique of Dr. Zuger's Manuscript," *Psychosomatic Medicine* 32 (1970): 463–65, 464.

34. Carey Roberts, "Death of a Gender Myth" (letter), *Washington Times*, May 23, 2004, section B, 5.

35. Mark Cochrane, "Boy/Girl/Boy," *Vancouver Sun*, March 5, 2000, n.p., via http://www.lexisnexis.com/uk/nexis. Our italics.

36. David Jones, "The Real Boy Called 'It,'" *Daily Mail*, May 15, 2004, 30–32, 32.

37. John Bancroft, "John Money: Some Comments on His Early Work," in *John Money: A Tribute*, ed. Eli Coleman (Binghamton, NY: Haworth, 1991), 1–8, 6.

38. Money and Ehrhardt, *Man and Woman*, xii.

39. Ehrhardt, "John Money, Ph.D.," 224.

40. Green, "John Money, Ph.D.," 631.

41. Paul R. Abramson, "For The Sake of Money: A Pre-Birthday Greeting to John Money," *Journal of Sex Research* 28 (1991): 1.

42. John Money, "Sex Hormones and Other Variables in Human Eroticism," in *Sex and Internal Secretions*, ed. William C. Young, 3rd ed., vol. 2 (Baltimore: Williams and Wilkins, 1961), 1383–1400, 1396.

43. Milton Diamond, "A Critical Evaluation of the Ontogeny of Human Sexual Behavior," *Quarterly Review of Biology* 40 (1965): 147–75, 162.

44. John Money, review of *The Intersexual Disorders* by Christopher J. Dewhurst and Ronald R. Gordon, *Journal of Nervous and Mental Disease* 152 (1971): 216–18, 217.

45. John Money, "Prenatal Hormones and Postnatal Socialization in Gender Identity Differentiation," in *Nebraska Symposium on Motivation*, ed. James K. Cole and Richard Dienstbier (Lincoln: University of Nebraska Press, 1973), 221–95, 232.

46. Money, *Sin, Science, and the Sex Police*, 314.

47. Linda Nicholson, "Interpreting Gender," *Signs* 20 (1994): 79–105, 82.

48. Money, "Gender: History, Theory," 76.

49. John Money and Margaret Lamacz, *Vandalized Lovemaps: Paraphilic Outcomes in Seven Cases of Pediatric Sexology* (Buffalo, NY: Prometheus, 1989), dedication, n.p.

50. Money, *Lovemaps*, 4.

51. Judith Halberstam, *The Queer Art of Failure* (Durham, NC: Duke University Press, 2011), 124.

52. Money and Ehrhardt, *Man and Woman*, 228.

53. John Money, "Hermaphroditism, Gender, and Precocity in Hyperadrenocorticism: Psychologic Findings," *Bulletin of the Johns Hopkins Hospital* 96 (1955): 253–64, 258.

# * 1 *

# *Mapping*

CHAPTER 1

# The Matter of Gender

*Nikki Sullivan*

> The social history of our era cannot be written without naming gender, gender role, and gender identity as organizing principles.
>
> JOHN MONEY, "The Concept of Gender Identity Disorder in Childhood and Adolescence after 39 Years"

In the popular imaginary of the present, John Money is most often cast as the quintessential social constructionist; as someone who claimed that gender is solely an effect of enculturation[1] and, as such, is radically mutable and alterable. For some, Money's purported theory of gender as an effect of nurture (as opposed to nature), made him "one of the gurus of the [second-wave] feminist movement"[2]—a characterization that seems to sit uncomfortably with Money's criticism of what he saw as feminism's "conceptual neutering of gender."[3] For others, Money's clinical attempts to "prove" his theory have—particularly in light of the David Reimer case—shown his ideas to be both flawed and dangerous, and his practice to be unethical. Interestingly, in one of the most influential popular cultural texts on Money, *As Nature Made Him: The Boy Who Was Raised as a Girl*, John Colapinto's characterization of Money as a constructionist is an inferential effect of the author's narrativization of a world of goodies and baddies, truth and lies, fact and fiction, males and females, rather than something that is convincingly argued through a close engagement with Money's writings. But despite this, Colapinto's text, and in particular his view of Money as a constructionist, seems largely to be taken as gospel.

Money himself would dispute this view of his work, arguing instead that his account of gender identity/role (G-I/R)—a concept I will discuss in detail in due course—is interactionist, that is, it acknowledges the generative effects of both biology and culture. Indeed, it is on this basis that Jennifer Germon has recently argued that "Money's gender offers a third wave of productive potential, one that differs from the second (as in second-wave feminism), precisely because his theories presuppose an interactive relation of cells to environment and to experience(s)."[4] Germon's optimism is challenged, however, by Lesley Rogers and Joan Walsh's much earlier claim that "it is not an interactionist approach to swing towards biological determinism most of the time and then occasionally, when it suits, to swing towards the environmental side."[5] As they see it, insofar as the model of gender attribution that Money and his coauthors articulate is underpinned by extremely conventional assumptions about and attitudes toward sexual difference, any attempt to articulate thoroughly the role of the social in the attribution of gender is wholly undermined. They write: "While Money, Ehrhardt, and co-workers consider the social aspects of gender . . . they take for granted that there are two genders, that there are differences between them, and that fundamentally gender is a consequence of a biological blueprint for behavior as well as physique."[6] Similarly, Ruth Doell and Helen Longino[7] have argued that Money's model of gender is more accurately additive than interactionist since it does not explain how biological and social variables work in tandem,[8] but rather, posits the biological as foundational.

There is little doubt that Money's work—in particular his elaboration of "gender"—has been hugely influential, and that rather than being confined to the worlds of scientific research and/or clinical practice, its influence has shaped us all. Consider, for a moment, the extent to which "gender" (however one might conceive it) has become central to everyday life, so much so that it is difficult to imagine how we might function without such a concept. This alone, it seems to me, is reason enough to engage with Money's vast œuvre. But further incentive comes from the fact that while competing interpretations of Money's work are readily available to those who choose to seek them out, the popular image of Money as a constructionist abounds. This characterization is an oversimplification of Money's work, and one which can only be maintained through a lack of engagement with his writing. If, as feminists have long argued, ongoing analyses of identity and difference—and in particular, so-called sexual difference—are politically imperative, then a close engagement with Money's highly influential texts, the assumptions that informed his claims, and the ongoing and multifarious effects such claims produce, likewise seems called for. In critically interrogating Money's account of G-I/R, my aim in this chapter is not so much to definitively classify his work as either constructionist or determinist, but rather, to trouble the very tendency to see in dimorphic terms since, as Helen Longino has noted, "as long as dimorphism remains at the centre of discourse, other patterns of difference remain hidden both as possibility and as reality."[9]

## MONEY'S "GENDER"

In popular parlance, gender tends either to be used as interchangeable with "sex" or, alternately, to refer to the social (as opposed to the so-called biological or "sexed") aspects of femaleness and maleness. In both cases, gender is a term that is commonly used to classify others and to refer to our own sense of ourselves as male or female, men or women, neither or both. Despite the fact that such conceptions of gender feel self-evident, they are, in fact, relatively recent. It has been claimed by Money and others that the first use of the term "gender" to refer to something other than feminine and masculine forms within language occurred in Money's 1955 publications "Hermaphroditism, Gender and Precocity in Hyperadrenocortism: Psychologic Findings" and, coauthored with Joan and John Hampson, "Hermaphroditism: Recommendations Concerning Assignment of Sex, Change of Sex, and Psychologic Management."[10] Indeed, it was through his early work with intersex patients that Money came to consider the term "sex" inadequate to describe the lived embodiment of those whose anatomies are either "discordant" or do not appear to match the sex roles associated with masculinity or femininity, and/or the sense of self a particular individual has.[11] For Money, intersexuality challenges the "commonsense" idea that sex (as naturally dichotomous) is a biological characteristic that at once determines genital morphology and can be determined with reference to that morphology, and that sex roles naturally follow from genital morphology and are concordant with it. It is possible, writes Money, "to have the genetical sex of a male . . . ; the gonadal sex of a male; the internal morphologic sex of a male; the external genital morphologic sex of a female; the hormonal pubertal sex of a female; the assigned sex of a female; and the gender-role and identity of a female."[12] Hence, Money's coining of the term "gender"—or, more precisely, gender identity/role (G-I/R)—to refer to the multivariate character of the "totality of masculinity/femininity, genital sex included,"[13] that each person attains even when the multiple aspects of the self (as a "man" or a "woman") are (seemingly) discordant. What we see here, then, is that for Money gender is not synonymous with "sex" (as a set of biological variables), but nor are the five aspects of sex that he identifies in the above quote entirely separable from G-I/R.[14]

Money conceives gender identity and gender role as "obverse sides of the same coin. They constitute a unity."[15] Without this unity, he argues, "gender role . . . become[s] a socially transmitted acquisition, divorced from the biology of sex and the brain"; it becomes "desexualized," "cleaned up."[16] As Money understood it, gender identity is the experience one has of oneself as a man or a woman—"the kingpin of your identity"[17] as he and Patricia Tucker describe it—and gender role is the manifestation of this sense of self in one's daily performance of self. In turn, one's performance of gender reaffirms one's gender identity, in particular because it is through gender role, as "everything that [one] says or does to indicate to others . . . the degree that one is either male, or female, or ambivalent,"[18] that others perceive and position one as gendered (in a specific way). There are, of course, situations in which others' perception of the gender of an individual may not fit with that individual's self-perception, but this is the exception rather than the rule, and it is something I will discuss in more detail in chapter 5.

As Money explains it, gender role is performative in two senses: it is an action or set of actions one articulates corporeally in a world of and with others, and, at the same time, it is constitutive of the self. In other words, gender role makes one be(come) male, female, neither or both, in and through what we might call—although Money himself does not use this term—sedimentation: the more we repeat certain actions, the more naturalized or habituated[19] such actions become, and the more they come to appear (both to others and to ourselves) as external expressions of who we "really" are. Clearly, then, while G-I/R may be effected by, for example, gonadal morphology or hormonal activity, it is never wholly determined by what we commonly think of as "biology." But nor, if G-I/R is intercorporeally (re)produced, can it be radically open and/or free, or, at least, that is what one might suppose. For Money, however, the story is a little more confused and confusing as we shall see.

For Money, the ability to acquire a G-I/R is "phylogenetically given, whereas the actuality is ontogenetically given."[20] In other words, while all humans share the ability to acquire a G-I/R, the G-I/Rs we each develop will differ according to context, pre- and postnatal history, morphology, and so on. For Money, then, the ability to acquire a G-I/R—which he refers to as a "phylism"[21]—is, like the ability to acquire language, to breathe, to laugh, or to "pairbond,"[22] sex-shared: he writes, "You were wired but not programmed for gender in the same sense that you were wired but not programmed for language."[23] There are, however, phylisms which, according to Money's conceptual schema, are exclusive to either men or women: these are lactation, menstruation, ovulation, and gestation, which are (allegedly) exclusive to women, and impregnation, which is (allegedly) exclusive to men. Money refers to these as "sex irreducible" dimensions of G-I/R[24] and differentiates them, in kind, from what he classifies as the sex derivative,[25] sex adjunctive,[26] and sex arbitrary or sex adventitious[27] dimensions of G-I/R. What begins to emerge here, then, is a categorical distinction between aspects of G-I/R that are universal and somehow determined by sex (as a set of biological variables)[28] and aspects that are context specific and an effect of a particular "society's customary way of doing things."[29] So, for example, while one can acquire the ability to operate a heavy goods vehicle in and through particular cultural processes (e.g., driving lessons), one cannot acquire the ability to ovulate, ontogenetically.[30] A similar ontological move is apparent in Money's claim that phylisms that are sex-shared are sometimes "threshold dimorphic,"[31] such that, for example, adolescent boys are more readily aroused by "sexy pin-up pictures"[32] than are adolescent girls. It is possible, writes Money:

> that divergent threshold levels are preset as early as in prenatal life when steroidal sex hormones organize bipotential brain regions and pathways to differentiate as predominantly either male or female. From animal experiments, there is abundant evidence that such organization does indeed take place.[33]

Throughout Money's work, the lowering of thresholds is vaguely associated with, although never convincingly connected—at least not in a straightforward causal sense—to prenatal exposure to hormones, most particularly androgen.[34] This is an issue I will engage with at length in chapter 5, but for now I want to suggest that in both the examples discussed we see that for Money G-I/R is never a purely social phenomenon and, while the acquisition of G-I/R necessarily involves what we might ordinarily think of as biological processes, these processes are never wholly determinative. For example, even though the ability to menstruate is associated with women, it does not guarantee a female G-I/R, as the existence of trans men or FTM (female-to-male) transsexuals shows. Conversely, an inability to menstruate does not mean that a person raised as female will not continue to identify and live as a woman once "amenorrhea" becomes apparent. Consequently, Money describes his model of G-I/R acquisition as biosocial[35] or interactionist, and as developmental (and sequential) as opposed to causal.[36]

As I said earlier, Money's proposition that identity does not strictly follow from anatomy, or, even when it appears to, that anatomy is not the *cause* of G-I/R, is an outcome of his early work with intersexuals, and later, with individuals desiring to undergo "sex reassignment" procedures. The clinical challenge Money faced in both cases was the question of whether or not surgery should be performed, and if so, on what basis its practice might be justified. Central to Money's theory of G-I/R, and to his recommendation that both infants with atypically sexed bodies and adults whose gender identity is (according to normative logic) at odds with their genital morphology should undergo surgical (trans)formation, is his concept of "the critical period of development."[37] Critical periods occur, according to Money, both anatomically and in the development of G-I/R more broadly, and each is marked, metaphorically speaking, by a gate that, once closed, is at best unlikely and at worst impossible to reopen.[38] This closing of gates along a developmental pathway "locks in" G-I/R such that one's sense of self and one's gendered performance becomes increasingly sedimented as one's subjectivity develops: you acquire, developmentally, a "native gender,"[39] or, to put it somewhat differently, "bipotentiality becomes monopotentiality."[40] Money writes:

> As you approached each gate's sex-differentiation point, you could have gone in either direction, but as you passed through, the gate locked, fixing the prior period of development as male or female. Your gonads, for example, could have become either testicles or ovaries, but once they became testicles they lost the option of becoming ovaries. . . . In behavior . . . at first you drove all over the highway, but as you proceeded you tended to stick more and more to the lanes marked out and socially prescribed for your sex.[41]

In *Sexual Signatures*, Money and Tucker cite what later became known as the Reimer case (discussed in detail in chapter 3) as "convincing evidence that the gender identity gate was wide open when you were born and stayed open for some time thereafter"[42] and further argue that transsexuals demonstrate that "the gate [is] also open for those who [are] sexually normal at birth"[43] and that "the gender identity gate locks tight once it closes."[44] This conception of the postnatal critical period in which G-I/R becomes "nativized" leads Money to argue in support of sex reassignment surgery (SRS), since the (developmentally "mature") transsexual's sense of self (as male or female) no longer retains the plasticity that the body's morphology will, to some extent, always have (since it is the physical nature of the body to change at the very least at the level of appearance).[45] Money and Tucker write, "Never yet has even the full weight of societal pressure, abetted by intensive psychotherapy, been able to reverse the gender identity of a trans[s]exual after it ha[s] differentiated completely."[46] This irreversibility, supported as it is by examples of patients whose intersex status did not become apparent until puberty but who refused to consider reassignment when "their gender-identities were challenged by their [changing] bodies,"[47] leads Money and Tucker to conclude that

> the importance of gender identity and the hazards of trying to change it once it has differentiated, make it vitally important to pinpoint the critical period for this stage of development. . . . Like other stages of development, this one varies somewhat with the individual.[48] In addition, cultural patterns of child training sometimes obscure the gender differentiation process. We can now say, however, that the critical period for gender identity differentiation coincides with the critical period for learning language.[49]

## GENDER IDENTITY DIFFERENTIATION AND/OR GENDERMAPS

Sexology, writes Money, "is the science of . . . the differentiation and dimorphism of sex,"[50] of the biosocial processes that shape us as women, men, and, very occasionally, androgynous. Sexology's advantage in the mapping of these processes lies in its multi- or transdisciplinary approach—an approach that, as Money tells it, allows him to avoid the all-too-common tendency to divorce the biological from the social, nature from nurture, sex from gender, body from mind. Indeed, from Money's perspective, the splitting of sex and gender and the establishment and maintenance of disciplinary boundaries and discipline-specific knowledges and practices have, historically, been mutually constitutive,[51] and have resulted in what he describes as "enormous doctrinal carnage."[52] Money asserts that his work bridges the chasm between social and biological definitions of gender, as well as that between the "hard" sciences and the social sciences in and through the conception of G-I/R as the effect of multivariate and sequential differentiation,[53] or gendermapping. Gendermap, writes Money,

> Is the term used to refer to the entity, template, or schema within the mind and brain (mindbrain) unity that codes masculinity and femininity and androgyny. . . . The gendermap is a conceptual entity under which are assembled all the male/female differences, and similarities also, not only those that are procreative and phylogenetically determined, but also those that are arbitrary and conventionally determined, such as male/female differences in education, vocation, and recreation. The gendermap . . . ha[s] a history of growth and development from very simple beginnings to very

> complex outcomes. [It is] multivariately and sequentially determined and, therefore, complicated to study. Explanations of [its] genesis are in terms of temporal sequences, not causal sequences.[54]

Figure 1 illustrates Money's conception of the temporal sequence (or gated forks along a single developmental pathway) integral to G-I/R acquisition,[55] "beginning with genetic foundations and terminating with social learning."[56]

Underpinning this model are two assumptions that are ripe for critical attention. The first is that there is a "normal" developmental teleology that, when followed in an orderly manner, will result in an ideal(ized), unified G-I/R. Consequently, gender-variant corporealities (or "gender identity errors" as Money sometimes calls them)[57] are conceived (and thus constituted) as the result of developmental deviations in the course of this natural(ized) universal system. Second, and related, is the assumption that G-I/Rs and the processes by which G-I/R differentiation occurs are naturally dimorphic and therefore naturally heterosexual.[58] In *Gay, Straight, and In-Between*, Money writes, "Gender coding is by definition dualistic. One half of the code is for female, the other for male. A child must assimilate both halves of the code, identifying with one and complementating the other."[59] Postnatally, "the child becomes conditioned to adhere to the positive model which is the one congruous with his rearing and, in the normal course of events, consistent with his anatomy. The opposite or negative-valence model becomes a constant reminder of how one should not act."[60] This process of coding by which two gender schemas develop in the "mindbrain" means that gendermaps are never exclusively social or biological in origin. Indeed, since learned behavior shapes the "mindbrain" ontogenetically (by building on what is phylogenetically laid down), it is both wrongheaded and impossible, argues Money, to attempt to differentiate the gendermaps in terms of social and biological aspects. "It is there, in the brain," writes Money, "that ontogeny and phylogeny meet; there that the social customs and traditions . . . are assimilated and fused with one's species heritage. In the brain sociopsycho-physiosomatic and/or somatophysiopsychosocial are one."[61] In short, then, gendermaps consist of "features that are phylogenetically shared with other members of the species, and characteristics that are ontogenetically personal,"[62] such that gendered subjectivity is at once unique, situated, and in-process, and shared, and partially (and increasingly) determined.

While this model may be useful insofar as it appears to enable a move beyond the limited and limiting choice between biological determinism and social constructionism, its productive potential is limited by (at least) two

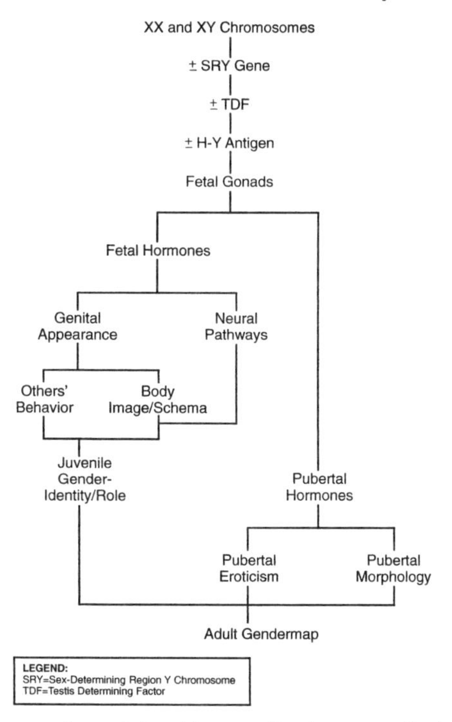


FIGURE 1. Diagram from John Money, *Gendermaps: Social Constructionism, Feminism, and Sexosophical History* (New York: Continuum, 1995), 98.

aspects that are integral to it. The first, which I will discuss in detail in chapter 5, is the association of "sedimentation" with the brain (or "mindbrain") and the ensuing priority given to the "brain" as the sort of "black box"[63] or control center of behavior and identity: the centrum, as Money calls it in an account of erotic behavior, from which "the genitalia, the organs of copulation, have their governance."[64] Elsewhere he writes,

> In the ultimate analysis, gender differences reside in the brain, regardless of how they got there, whether as a consequence of prenatal or neonatal hormonal programming, postnatal social programming, pubertal hormonal programming, or geriatric deterioration of the programming from earlier years. . . . Let it never be forgotten that learning itself is represented in the brain. There is, indeed, a brain biology of learning, irrespective of our ignorance of it.[65]

As vague and unsatisfying as this may sound to some readers, the ("neurosexist") association of sexual difference (as "naturally dimorphic") with "the brain" is, as Cordelia Fine has shown, by no means new, nor does it show any sign of diminishing in popularity.[66]

The second (related) problem with Money's model is the presupposition of dimorphism. At no stage in Money's vast œuvre does he make any attempt to substantiate his conception of G-I/R as "naturally dimorphic." Instead, it is simply posited as a given, as being self-evident and without question. Given the far-reaching effects of this assumption not only on Money's own work, but more particularly, on commonsense ideas about difference, the natural, deviance, and so on, I want to ponder for a moment what exactly the conceptual scaffolding might be that makes the dimorphic model of G-I/R so compelling (both to Money, and, it seems, to a large proportion of contemporary Western society more generally). The answer, I contend, lies in a relatively small but illuminating passage in *Sexual Signatures* in which Money and Tucker assert that

> The irreducible requirement for the survival of humanity is that men and women cooperate *as* men and women at least well enough to survive, reproduce, and rear a new generation. A man's ability to impregnate and a woman's to menstruate, gestate, and lactate, are not, by themselves an adequate basis for cooperation. . . . Gender stereotypes, with all their many more or less arbitrary sex distinctions, provide the framework for that cooperation. They must start from the four basic reproductive functions, but they cannot end there.[67]

Later in the same text, this naturalized and naturalizing notion of complementation is reiterated in a discussion of "dating and mating" as socially regulated, but primarily "matters of the brain and hormones, best understood in the light of evolutionary development."[68] I want to suggest that this is clearly a case of "extrascientific"[69] considerations (that is, preconceived ideas and values) shaping "scientific" inquiry.[70] As Anne Fausto-Sterling has convincingly argued, "scientists in analyzing male/female differences peer through the prism of everyday culture. More often than not their hidden agendas that are unarticulated bear strong resemblances to broader social agendas"[71]—a claim I will elaborate in the discussion of the phenomenological understanding of orientation in chapter 5. It is also an example of the legitimation of both gender stereotypes and what Judith Butler refers to as "the heterosexual matrix,"[72] through "pseudo-scientific" explanations in which, as Fine notes, the word "brain" works as if by magic to render such claims "modern and scientific rather than old-fashioned and sexist."[73]

As Butler has argued, the heterosexual matrix is necessarily exclusionary; it operates such that one term can only gain its identity and its position as "normal," as "subject," as male, and so on, through the abjection of that which it allegedly is not. And while Money recognizes this logic insofar as it operates to normalize male(ness) and female(ness) as polarities—as that to which the individual adheres or from which he or she distances him- or herself in and through the processes of identification and complementation—he does not acknowledge that another, less palatable, effect of such logic is to constitute "gender-variant" corporealities as necessarily abject(ed)—an effect that seems uncomfortably at odds with the "libertarian Money's" claim that homosexuality (as an erotic orientation that is never simply freely chosen)[74] is not properly paraphilic,[75] and should be accepted by society as a "viable option" that should therefore not be subject to enforced medical treatment.[76] Why this should be the case, when intersexuality clearly, in Money's eyes, calls for medical intervention is, ultimately, unclear. In short, what Money fails to explicate is the fact that on his model, cisgender and heterosexuality can only be (conceived/constituted as) the "proper" outcomes of a natural(ized) process of development precisely if (and because) "gender discordance"—or genderqueer—and "nonheterosexuality" are simultaneously produced as abject(ed). If this is the case, then cisgender and "gender discordance," heterosexuality and homosexuality, are less the polarized effects of processes that are first and fundamentally biological (even if this biological foundation is then supplemented by social learning) than of processes that are social and discursive and that have real material effects, that is, that shape bodily-being-in-the-world in the most profound (and, in the case of intersexuality, literal) ways. In disavowing (or at least ignoring) the fact that the abjected other "is, after all, 'inside' the subject as its own founding repudiation,"[77] Money fails to acknowledge the debt to the other that is incurred in the articulation and taking up of the position of psychosexual normalcy. In chapter 5, I will return to this critique of the assumption of dimorphism in Money's work and explore the ways in which such an assumption intersects with and is supported by Money's turn to the brain or, more specifically, to "brain organization theory."

## MONEY'S GENDER AND THE QUESTION OF INFLUENCE

At the beginning of this chapter, I suggested that Money is commonly regarded in the contemporary context as a social constructionist—a characterization he himself would strongly refute. I also noted that as such, some have claimed him as the darling of second-wave feminism. Money's model of G-I/R outlined here clearly shows that insofar as the former claim is misguided, the latter too is unlikely to carry any real weight. Indeed, Money's criticism of second-wave feminism's alleged divorcing of sex from gender, combined with the accusation of essentialism made by the very small number of feminists who engaged with Money's work in the period associated with feminism's so-called second wave, clearly points to the dubiousness of such claims.[78] Having said this, however, I do not mean to imply that second-wave feminists embraced a model of gender that was simply opposed to Money's: to do so would be to reaffirm the overly simplistic dichotomy of social constructionism versus biological determinism. What I want to suggest instead is that while at the time of its development Money's model of G-I/R had little impact on second-wave feminism(s),[79] it does, ironically—and this is a claim that may well make Money turn in his grave—share with many second-wave feminist accounts of gender an unacknowledged dependence on "the sexed body" as somehow foundational.[80] In short, it is my contention that both Money's work on gender and the work of the vast majority of second-wave feminists exemplifies what Linda Nicholson refers to as "biological foundationalism."[81]

Nicholson explains this term with reference to what she describes as the "coatrack view of self-identity."[82] On this model, the rack is the aspect of the self that is common to everyone (or, in the case of feminist writing, to all women); it is the thing on which the cultural aspects of selfhood—the things that make us different—are hung. Money's coatrack is what he calls "the phylogenetic," and what gets hung on that foundation are the ontogenetic aspects of G-I/R. This model is also apparent in the work of diverse thinkers associated with second-wave feminism. For example, anthropologist Gayle Rubin introduced the phrase "the sex/gender system"[83] in her landmark essay "The Traffic in Women" (1975), to refer to "a set of arrangements by which the biological raw material of human sex and procreation is shaped by human, social intervention."[84] Feminism, Rubin argued, should aim to create a "genderless (though not sexless) society, in which one's sexual anatomy is irrelevant to who one is, what one does, and with whom one makes love."[85]

In a very different feminist project from the same era, Janice Raymond vehemently attacked MTF (male-to-female) transsexuals, in particular those who identified as lesbian, arguing that "male-to-constructed-female" lesbian feminists "can only play the part"[86] of women, lesbians, and feminists. Further, she claimed that

> All [MTF] transsexuals rape women's bodies by reducing the real female form to an artifact, appropriating this body for themselves. However, the transsexually constructed lesbian-feminist violates women's sexuality and spirit as well. . . . Because [MTF] transsexuals have lost their physical "members" does not mean that they have lost their ability to penetrate women—women's minds, women's sexuality, [MTF] transsexuals merely cut off the most obvious means of invading women so that they seem non-invasive. However, as Mary Daly has remarked in the case of the transsexually constructed lesbian-feminists their whole presence becomes a "member" invading women's presence and dividing us once more from each other.[87]

While I do not want to claim that Raymond explicitly posits a distinction between sex and gender, nature and culture in her work, the criticisms she aims at MTFs clearly imply the existence of a sexuality, a spirit, a body, that is particular to (all) women and that is not, by definition, shared by men. And even though Raymond recognizes differences between men—for example, not all men want to be transsexuals—it nevertheless seems that fundamentally all men are the same at the very least insofar as they are not, and cannot ever be, women.

This sense of sex-shared character(istics) is also apparent in Carol Gilligan's (1983) work on what she perceives as a universal proclivity among women for "relatedness."[88] While Gilligan, like Raymond, describes this tendency as an aspect of "women's culture"—that is, as something that occurs in and through social processes and organization rather than solely as a result of biology—there is a sense in which both writers, "claim a strong [universal(izing)] correlation between people with certain biological characteristics [that is, penises or vaginas] and people with certain character traits."[89] As Nicholson notes, claims that women are different from men (or vice versa) in "such and such ways,"[90] ultimately function to constitute certain characteristics as female or male, and in doing so, to naturalize the differences purportedly described. These differences, Nicholson claims, "tend to reflect the perspective of those making the characterizations . . . and to reflect the biases"[91] of particular social groups, most often those in a position of power and/or authority[92]—a claim I will return to in a moment.

In pointing out what I, following Nicholson, identify as the biological foundationalism of these very different projects, it is not my intention to conflate them, and in so doing, to flatten out the important theoretical and political differences between them. Indeed, what I find so useful about this concept is that it enables a move beyond the black-and-white logic of the essentialism versus constructionism debate: it allows, indeed it calls for, an analysis of the nuances of particular positions, and the (similar and different) effects they produce. As Nicholson puts it, "Biological foundationalism is not equivalent to biological determinism; all of its forms, though some more extensively so than others, include some element of social construction."[93] Biological foundationalism "is best understood as representing a continuum of positions. . . . [This] counters a commonplace contemporary tendency to think of social constructionist positions as all alike in the role that biology plays within them."[94] Given this, the question is not so much whether biological foundationalism is a good thing or a bad thing since clearly it generates heterogeneous effects even within one particular perspective. Nor is it a question of how to best clarify gender, or of whose/which conception is most correct. If, as Nicholson claims, "the clarification of the meaning of . . . any concept . . . is stipulative"[95] (rather than merely describing a given reality), then it is so within the context of a particular regime of truth, within a particular discipline and/or set of disciplinary practices.[96] Our conceptions of gender are, to quote Nicholson again, situated: "they emerge from our own places within history and culture; they are political acts that reflect the contexts we emerge out of and the futures we would like to see."[97] Nowhere is this clearer than in Money's work on transsexualism in which he simultaneously claims to know of "no proven cause of gender identity disorder,"[98] and yet turns to studies on animals to explain how hormonal coding can affect the G-I/R of sheep who have been androgenized and thereafter try to mount other ewes and are not approached by rams;[99] androgenized female monkeys who "played more boisterously than normal female monkeys" and whose "assertiveness and mating behavior . . . fell somewhere in between the normal male and female";[100] and of developmentally "normal" white leghorn roosters whose sexual attraction to and mounting of a headless model of a chicken demonstrates "the fact that [non-cross-coded] men's erotic arousal, including erection of the penis, is eye-sensitive and can be rather easily triggered by a visual stimuli."[101]

To a humanities scholar, these claims, and the experiments that make them possible, may appear extremely spurious, and yet in the context in which Money carried out his research, it seems they were not. Moreover, the fact that experiments on animals continue to be carried out (in huge numbers, and often in ways that are, to many, inhumane) in the hope that they will one day provide the key(s) to the difference(s) between the sexes is both mind-boggling and profoundly telling in terms of the investment contemporary cultures clearly have in the difference they seem so desperately to desire. If, as Nelson Goodman has stated, "scientific development always starts from worlds already on hand,"[102] then one wonders what worlds shaped Money's ideas and his practice. While a definitive answer to this question is, no doubt, impossible, I want to suggest that one such world is the field of sex endocrinology (including, and perhaps particularly, its uptake by biologists and zoologists), and, more particularly, what has come to be known as brain organization theory.[103] As a consequence, Money's work has tended to find its biggest influence not, as some have suggested, in feminism, or even in politicized accounts of gendered selfhood more generally, but rather, in brain organization research, and in the (largely clinical) discourses and practices surrounding intersexuality and transsexualism as I will demonstrate in chapter 5.

## NOTES

1. See, for example, John M. Sloop, *Disciplining Gender: Rhetorics of Sex Identity in Contemporary U.S. Culture* (Amherst: University of Massachusetts Press, 2004); Lynn Conway, "Basic TG/TS/IS Information," available at http://ai.eecs.umich.edu/people/conway/TS/TS.html, accessed November 14, 2011; Maggie McNeill, "Social Construction of Eunuchs" (2011), available at http://maggiemcneill.wordpress.com/2011/07/18/social-construction-of-eunuchs, accessed November 10, 2011. In a brief discussion of Harry Benjamin's *The Transsexual Phenomenon*, in an article on *TS Si*—a website "dedicated to the acceptance, medical treatment and legal protection of individuals correcting the misalignment of their brains and their anatomical sex, while supporting their transition into society as hormonally reconstituted and surgically corrected citizens"—the unnamed author claims that Benjamin's approach "ran directly counter to the nurture-oriented claims of John Money"; available at http://ts-si.org/horizons/1407-background-on-dr-john-money, accessed November 10, 2011. See also Eliserh, "Some Reflections on the State of Gender: Nature *vs.* Nurture" (2005), available at http://ts-si.org/horizons/1407-background-on-dr-john-money, accessed November 29, 2011; Sophia Siedlberg, "I'm Not Dissing You" (2008), available at http://www.intersexualite.org/Dissing.html, accessed November 20, 2011; Anna Dela Cruz, "Gender Self-Identity among Males: A Case for Biology" (2009), available at http://serendip.brynmawr.edu/exchange/node/4362, accessed November 12, 2011; William Reville, "Gender Can't Be Freely Chosen," *Irish Times* (2011), available at http://www.irishtimes.com/newspaper/sciencetoday/2011/0616/1224298993467.html, accessed June 20, 2011; Myria, "From the Forum: Constructed Gender?" (2003), available at http://www.ifeminists.net/introduction/editorials/2003/0520myria.html, accessed July 20, 2010; Hanna Rosin, "A Boy's Life," *Atlantic* (2008), available at http://www.theatlantic.com/magazine/archive/2008/11/a-boy-apos-s-life/7059/5/, accessed August 10, 2011; Wendy Cealey Harrison, "The Shadow and the Substance: The Sex/Gender Debate," in *Handbook of Gender and Women's Studies*, ed. K. Davis, M. Evans and J. Lorber (London: Sage, 2006), 35–52;

James H. Liu, "Sexual Assignment and Management of the Transsexual Individual," *Journal of Family Practice* 7 (2009), available at http://www.jfponline.com/Pages.asp?AID=7352, accessed March 5, 2011.

2. Babette Francis, "Is Gender a Social Construct or a Biological Imperative?," 2000, available at http://www.endeavourforum.org.au/articles/babette_social.html, accessed November 14, 2011. See also Schala's post on the blog *Gender Liberation beyond Feminism*, available at http://www.pellebilling.com/2009/03/gender-and-biology, accessed November 10, 2011.

3. See John Money, "The Conceptual Neutering of Gender and the Criminalization of Sex," *Archives of Sexual Behavior* 14 (1985): 279–90; John Money, *Gendermaps: Social Constructionism, Feminism, and Sexosophical History* (New York: Continuum, 1995), 72–73; and John Money, *Venuses Penuses: Sexology, Sexosophy and Exigency Theory* (New York: Prometheus, 1986), 591–600.

4. Jennifer E. Germon, *Gender: A Genealogy of an Idea* (New York: Palgrave Macmillan, 2009), 3.

5. Lesley Rogers and Joan Walsh, "Shortcomings of the Psychomedical Research of John Money and Co-Workers into Sex Differences in Behavior: Social and Political Implications," *Sex Roles* 8 (1982): 269–81, 278.

6. Rogers and Walsh, "Shortcomings," 272.

7. Ruth G. Doell and Helen E. Longino, "Sex Hormones and Human Behavior: A Critique of the Linear Model," *Journal of Homosexuality* 15 (1988): 55–78.

8. Rebecca M. Jordan-Young, *Brain Storm: The Flaws in the Science of Sex Differences* (Cambridge, MA: Harvard University Press, 2010), 8.

9. Helen E. Longino, *Science as Social Knowledge: Values and Objectivity in Scientific Inquiry* (Princeton: Princeton University Press, 1990), 171.

10. See, for example, John Money, "Gender: History, Theory and Usage of the Term in Sexology and Its Relationship to Nature/Nurture," *Journal of Sex and Marital Therapy* 11 (1985): 71–79, 71–72; Money, "Conceptual Neutering," 280; Money, *Gendermaps*, 18; John Money, "The Concept of Gender Identity Disorder in Childhood and Adolescence after 39 Years," *Journal of Sex and Marital Therapy* 20 (1994): 163–77, 163; Germon, *Gender*, 22.

11. In *Gay, Straight, and In-Between: The Sexology of Erotic Orientation* (Oxford: Oxford University Press, 1988), Money writes, "used strictly and correctly, gender is more inclusive than sex. It is an umbrella under which are sheltered all the different components of sex difference, including the sex-genital, sex-erotic, and sex-procreative components" (52–53).

12. Money, "Gender: History, Theory and Usage," 73.

13. Money, "Gender: History, Theory and Usage," 71.

14. Money, *Gendermaps*, 18–20.

15. Money, "Conceptual Neutering," 285. See also Money, "Gender: History, Theory and Usage," 71.

16. Money, "Conceptual Neutering," 275.

17. John Money and Patricia Tucker, *Sexual Signatures: On Being a Man or a Woman* (Boston: Little, Brown, 1975), 5.

18. Money, "Conceptual Neutering," 285. Interestingly, this formulation overlooks the possibility that one might feel certain that one is neither male or female.

19. In *Sexual Signatures*, Money and Tucker talk about how difficult it would be to simply assume a gender at will and/or to perform it "successfully" (i.e., so that one passes), 121–22. This is reminiscent of Butler's critique in *Bodies That Matter*, of the "wardrobe of gender" theory as a misreading of the notion of gender as performativity. See Judith Butler, *Bodies That Matter: On the Discursive Limits of "Sex"* (New York: Routledge, 1999).

20. Money, *Gendermaps*, 36.

21. Money uses the term "phylism" to refer to what he describes as "a unit or building block of our existence that belongs to us as individuals through our heritage as members of a species." See *Gendermaps*, 36.

22. For an account of Money's theory of "pairbonding" as a basic phenomenon of human existence, see John Money, "Sex, Love, and Commitment," *Journal of Sex and Marital Therapy* 2 (1976): 273–76.

23. Money and Tucker, *Sexual Signatures*, 89.

24. In "Destereotyping Sex Roles," *Society* 14, no. 5 (1977): 25–28, Money refers to the sex irreducible dimensions of G-I/R "sex roles," and differentiates them from "sex-coded roles"—a term he seems to use as an umbrella for the other dimensions of G-I/R listed above. In *Gendermaps* he speaks of impregnation, menstruation, gestation, and lactation as "biological imperatives that are laid down for all men and women" (38).

25. Urinary position is an example Money gives of sex-derivative behavior.

26. Money gives the example of the extension of (what he sees as) men's territorial roaming to truck driving. For a more detailed account of these dimensions of G-I/R see John Money, "Gender: History, Theory and Usage," 74. See also Money "Destereotyping Sex Roles," 26; and Money, *Gay, Straight and In-Between*, 54–70.

27. This includes things such as hairstyles, forms of adornment, etc.

28. Money does accede that what he names as "sex irreducible" differences between women and men can be affected by what we might think of as cultural influences during fetal development. For example, drugs given to pregnant women to lessen the risk of miscarriage have, in some cases, androgenized fetuses otherwise marked as "female" such that they are born intersexed. In some such cases, those raised as (androgenized) women are not able to ovulate, menstruate, or gestate. See, for example, John Money and Jean Dalery, "Iatrogenic Homosexuality: Gender Identity in Seven 46,XX Chromosomal Females with Hyperadrenal-cortical Hermaphroditism Born with a Penis, Three Reared as Boys, Four Reared as Girls," *Journal of Homosexuality* 1 (1976): 357–71. In "The Future of Sex and Gender," *Journal of Clinical Child Psychology* 9 (1980): 132–33, Money imagines a future in which "male-to-female transsexuals who yearn for a pregnancy may be able to rely on a . . . procedure of induced ectopic pregnancy." This would involve the transplantation of a fertilized egg into the peritoneal cavity such that it would implant itself on the wall of the bowel. He also adds that "whatever can be envisaged for transsexuals can be envisaged for nontranssexuals as well," thus implying the possibility that gestation may not always be the sole domain of women (132).

29. Money, *Gendermaps*, 51.

30. This sort of "commonsense" logic is debatable given that men have been known to lactate, and, more recently, trans men have become pregnant and given birth.

31. Money, *Gendermaps*, 37. See also Money, "Destereotyping Sex Roles," 27–28. In

*Gendermaps*, Money identifies nine parameters of sex-shared, threshold-dimorphic behavior patterns on which educational and vocational male/female differences are developmentally superimposed. These are general kinesis, competitive rivalry, roaming and territorial boundary mapping or marking, territorial defense, guarding of young, nesting and homemaking, parental care of young, sexual positioning, and erotic arousal. (See *Gendermaps*, 39–47.)

32. Money, *Gendermaps*, 37.

33. Money, *Gendermaps*, 38.

34. See, for example, *Sexual Signatures*, 67–78. See also John Money, "Gender-Transposition Theory and Homosexual Genesis," *Journal of Sex and Marital Therapy* 10 (1984): 75–82; and John Money, "The Influence of Hormones on Sexual Behavior," *Annual Review of Medicine* 16 (1965): 67–82.

35. Money, "Gender: History, Theory and Usage," 77.

36. John Money and Clay Primrose, "Sexual Dimorphism and Dissociation in the Psychology of Male Transsexuals," in *Transsexualism and Sex Reassignment*, ed. Richard Green and Money [1969] (Baltimore: Johns Hopkins University Press, 1975), 115–31, 127; Money and Tucker, *Sexual Signatures*, 6.

37. Money, "Gender: History, Theory and Usage," 74; Money, *Gendermaps*, 23 and 95; John Money, "Ablatio Penis: Normal Male Infant Sex-Reassignment as a Girl," *Archives of Sexual Behavior* 4 (1975): 65–71, 66–67. Early in his career Money used the concept of imprinting to refer to what he later calls the critical period.

38. Money, *Gendermaps*, 75–76. See also Money, "Future of Sex and Gender," 132, in which the author argues that "nature's plan is to resolve the original state of hermaphroditism or indeterminism once and for all, early in development, and from then on the sexual anatomy remains fixed." However, he goes on to envisage a future in which "humankind may unravel the secret of how nature goes about programming spontaneous sex reversal in some species and apply that secret to human beings." Indeed, he claims that "nature has already prepared the way" for "reverse embryology" in the figure of the transsexual.

39. Money and Tucker, *Sexual Signatures*, 89.

40. Money, *Gendermaps*, 111.

41. Money and Tucker, *Sexual Signatures*, 73.

42. Money and Tucker, *Sexual Signatures*, 90.

43. Money and Tucker, *Sexual Signatures*, 91.

44. Money and Tucker, *Sexual Signatures*, 98.

45. Money and Tucker assert that "people can no more be expected to decode behavior that has been locked into the core of their gender schemas than a Chinese woman whose feet were bound in childhood could be expected to walk naturally" (*Sexual Signatures*, 231).

46. Money and Tucker, *Sexual Signatures*, 98.

47. Money and Tucker, *Sexual Signatures*, 98.

48. Money suggests a variety of ages marking the point at which gender identity becomes sedimented. In *Sexual Signatures*, he and Tucker claim that it happens "before you got three, or at most, four, candles on your birthday cake" (119). In the same book, the authors suggest that the John/Joan case shows that "social forces can intervene decisively at least up to a year and a half after birth" (91).

49. Money and Tucker, *Sexual Signatures*, 100. Money and Tucker cite two case studies that they use to support this hypothesis. See 101–9.

50. John Money, "Sexology: Behavioral, Cultural, Hormonal, Neurological, Genetic, Etc.," *Journal of Sex Research* 9 (1973): 1–10, 10. See also John Money, "The Development of Sexology as a Discipline," *Journal of Sex Research* 12 (1976): 83–87, 86.

51. See Money, *Gendermaps*, 23–35, 72–76; Money, "Sexology: Behavioral, Cultural, Hormonal," 3–5; Money, "Gender: History, Theory and Usage," 77; Money, "Conceptual Neutering," 282–85.

52. Money, "Concept of Gender Identity Disorder," 168.

53. See Money, "Conceptual Neutering," 282–85.

54. Money, *Gendermaps*, 96–97. According to Money, the gendermap overlaps with the lovemap but neither is reducible to the other.

55. This is the most detailed example of the diagrammatic representations of gendermapping found throughout Money's publications.

56. Money and Primrose, "Sexual Dimorphism and Dissociation," 127. For a discussion of each of the multivariate sequential determinants identified in this diagram, see Money, *Gendermaps*, 97–108. See also *Sexual Signatures*, 41–50, for a detailed account of what Money identifies as major sex differentiation forks that occur prior to birth. These are (1) chromosomal differentiation at conception; (2) gonadal—six weeks after conception differentiation of ovaries/testes kicks in—Adam Principle; (3) fetal hormonal—Wolffian/Müllerian structures develop in accordance with hormone mix; (4) internal morphological; (5) external morphological and neural pathways. *Gendermaps*, 41–48.

57. John Money, "Sex Reassignment as Related to Hermaphroditism and Transsexualism," in *Transsexualism and Sex Reassignment*, ed. Richard Green and John Money (Baltimore: Johns Hopkins University Press, 1969), 91–113. Here Money discusses "errors" that result in intersexuality and transsexualism, but elsewhere configures homosexuality, bisexuality, and transvestism as examples of "gender transposition" effected by developmental errors in gendermapping. See John Money and Michael De Priest, "Three Cases of Genital Self-Surgery and Their Relationship to Transsexualism," *Journal of Sex Research* 12 (1976): 283–94, 292. See also "Sexual Dimorphism and Dissociation," 127, in which Money and Primrose use terms such as "maldevelopment" and "defect" to describe deviations from the developmental norm he posits in the diagram shown in figure 1.

58. In "The Concept of Gender Identity Disorder," Money writes, "The coding of language in the brain is bipolar. . . . The developmental coding of G-I/R in the brain is also bipolar" (173).

59. Money, *Gay, Straight, and In-Between*, 72. In order to avoid the kinds of criticisms posed by feminists and others of the association of masculinity with activeness and femininity with passivity in a dimorphic model of gender and heterosexuality, and to gesture toward some sort of "coital parity," Money adds the terms "quim" and "swive" to the analytic vocabulary of sexology to refer respectively to the "tak[ing of] the penis into the vagina and perform[ing], grasping, sliding, and rotating movements on it of varying rhythm, speed and intensity," and the "put[ting of] the penis into the vagina and perform[ing] sliding movements of varying depth, direction, rhythm, speed, and intensity." See John Money, "To Quim and

to Swive: Linguistic and Coital Parity, Male and Female," *Journal of Sex Research* 18 (1982): 173–76, 175.

60. Money and Primrose, "Sexual Dimorphism and Dissociation," 129.

61. John Money, "Sexosophy: A New Concept," *Journal of Sex Research* 18 (1982): 364–66, 366.

62. Money, *Gendermaps*, 104.

63. Doell and Longino use this analogy in their critique of what they see as Money's linear model of development. See "Sex Hormones and Human Behavior," 59.

64. Money, "Sexosophy," 365.

65. Money, "Conceptual Neutering," 285–86.

66. Cordelia Fine, *Delusions of Gender: The Real Science behind Sex Differences* (London: Icon Books, 2010). Part 2 of the book is entitled "Neurosexism," and it is from there that I borrow the phrase "neurosexist." Contemporary examples of "neurosexist" accounts of sexual difference include Louanne Brizendine, *The Female Brain* (London: Bantam Press, 2007); Simon Baron-Cohen, *The Essential Difference: Men, Women, and the Extreme Male Brain* (London: Allen Lane, 2003); Anne Moir and David Jessel, *Brain Sex: The Real Differences between Men and Women* (London: Michael Joseph, 1989); Allan Pease and Barbara Pease, *Why Men Don't Listen and Women Can't Read Maps: How We're Different, and What to Do about It* (New York: Welcome Rain, 2000); and Michael Gurian, *What Could He Be Thinking? A Guide to the Mysteries of a Man's Mind* (London: Element, 2004).

67. Money and Tucker, *Sexual Signatures*, 38–39.

68. Money and Tucker, *Sexual Signatures*, 170.

69. Nelly Oudshoorn, drawing on the work of Ludwik Fleck, uses the concept of "prescientific ideas" to emphasize the culturally shaped character of scientific knowledge. See Oudshoorn, *Beyond the Natural Body: An Archaeology of Sex Hormones* (London and New York: Routledge, 1994), 11.

70. For an insightful analysis of the ways in which this occurs, see Helen E. Longino and Ruth G. Doell, "Body, Bias, and Behavior: A Comparative Analysis of Reasoning in Two Areas of Biological Science," *Signs* 9 (1983): 206–27.

71. Anne Fausto-Sterling, *Myths of Gender: Biological Theories about Women and Men* (New York: Basic, 1985), 9.

72. Judith Butler, *Gender Trouble: Feminism and the Subversion of Identity* (New York: Routledge, 1990).

73. Fine, *Delusions of Gender*, 172.

74. In *Gay, Straight, and In-Between*, Money describes the claim that "heterosexuality and homosexuality have their origin in voluntary choice and are therefore already fully explained by fiat, without the superfluous addition of more research" as a "scientific fallacy" (85).

75. Money, *Gay, Straight, and In-Between*, 84. Money's claim, here, is somewhat ambiguous. He writes, "The classification of homosexuality as a paraphilia is scientifically untenable, insofar as all of the forty-odd paraphilias may occur in association with homosexual, heterosexual, or bisexual mating. Thus it is necessary to have a conceptual term other than paraphilia for the name of whatever it is that makes homosexual different from heterosexual." For further discussion of Money's conception of paraphilia, see chapters 2 and 6.

76. John Money, "Bisexual, Homosexual, and Heterosexual," *Journal of Homosexuality* 2 (1977): 229–33, 233.

77. Butler, *Bodies That Matter*, 237.

78. An exception to the generally held view among the few feminists who actually engaged with Money's work that it was mired in biological determinism can be found in Ann Oakley's *Sex, Gender, and Society* in which the author recognizes what Money describes as the interactionist character of his work. However, for Oakley, Money's analysis fails to articulate the political nature of gender, and the political import of such an insight. Consequently, she finds little reason to deploy his analysis. See *Sex, Gender and Society* (London: Temple Smith, 1972), 170.

79. One might claim that Money's work had an indirect influence inasmuch as it informed Robert Stoller's work on sex and gender, and this was taken up directly by some second-wave feminists. However, Stoller, unlike Money, makes a distinction between sex and gender. For more on Stoller and Money, see chapter 2.

80. My proposition is supported by Celia Roberts's claim that Money and Ehrhardt "positioned biological sex as a structuring materiality that interacts with culture to make gender." See *Messengers of Sex: Hormones, Biomedicine and Feminism* (Cambridge: Cambridge University Press, 2007), 5.

81. Linda Nicholson, "Interpreting Gender," *Signs* 20 (1994): 79–105, 80.

82. Nicholson, "Interpreting Gender," 81.

83. Gayle Rubin, "The Traffic in Women: Notes on the Political Economy of Sex," in *Toward an Anthology of Women*, ed. Rayna R. Reita (New York: Monthly Review, 1975), 157–210.

84. Rubin, "Traffic in Women," 165.

85. Rubin, "Traffic in Women," 204.

86. Janice Raymond, *The Transsexual Empire: The Making of the She-Male* (London: Women's Press, 1980), 103.

87. Raymond, *Transsexual Empire*, 104.

88. Carol Gilligan, *In a Different Voice: Psychological Theory and Women's Development* (Cambridge, MA: Harvard University Press, 1983). For critiques of the universalizing tendency in Gilligan's work, see John Broughton, "Women's Rationality and Men's Virtues," *Social Research* 50 (1983): 597–642; Linda Nicholson, "Women, Morality and History," *Social Research* 50 (1983): 514–36; Judy Auerbach, Linda Blum, Vicki Smith, and Christine Williams, "Commentary: On Gilligan's *In a Different Voice*," *Feminist Studies* 11 (1985): 149–61.

89. Nicholson, "Interpreting Gender," 94.

90. Nicholson, "Interpreting Gender," 94.

91. Nicholson, "Interpreting Gender," 94.

92. Raymond's work on MTF transsexuals—as fundamentally male, and fundamentally different from the women they allegedly wish to become—has been subject to severe criticism on this basis.

93. Nicholson, "Interpreting Gender," 92.

94. Nicholson, "Interpreting Gender," 88.

95. Nicholson, "Interpreting Gender," 102.

96. Foucault uses the term "regime of truth" to refer to "the types of discourse [a discipline]

accepts and makes function as true; the mechanisms and instances which enable one to distinguish true and false statements, the means by which each is sanctioned; the techniques and procedures accorded value in the acquisition of truth; the status of those who are charged with saying what counts as true." See Michel Foucault, "Truth and Power," in *Power/Knowledge: Selected Interviews and Writings, 1972–1977*, ed. Colin Gordon and trans. C. Gordon, L. Marshall, J. Mepham, and K. Soper (New York: Pantheon Books, 1980), 131.

97. Nicholson, "Interpreting Gender," 103.

98. Money, "Concept of Gender Identity Disorder," 170.

99. Money, "Concept of Gender Identity Disorder," 170–72.

100. Money and Tucker, *Sexual Signatures*, 67–8.

101. John Money and John G. Brennan, "Heterosexual vs. Homosexual Attitudes: Male Partners' Perception of Feminine Image of Male Transexuals," *Journal of Sex Research* 6 (1970): 193–209, 205.

102. Nelson Goodman, *Ways of Worldmaking* (Indianapolis: Hackett, 1978), 6.

103. For a detailed account of the emergence of organization theory in the 1950s, Money's contribution to it, and the historico-cultural reasons for its acceptance (in scientific circles), see Marianne Van Den Wijngaard, *Reinventing the Sexes: The Biomedical Construction of Femininity and Masculinity* [1991] (Bloomington: Indiana University Press, 1997), 27–46.

CHAPTER 2

# A Disavowed Inheritance: Nineteenth-Century Perversion Theory and John Money's "Paraphilia"

*Lisa Downing*

> It is possible that the West has not been capable of inventing any new pleasures, and it has doubtless not discovered any original vices. But it has defined new rules for the games of powers and pleasures. The frozen countenance (*visage figé*) of the perversions is a fixture of this game.
>
> MICHEL FOUCAULT, *The Will to Knowledge*, 48.

In this epigraph, Michel Foucault moves the focus of interrogation and investigation from "the sexual perversions," those stigmatized desires and practices that disobey the normative and utilitarian rules of sexuality and gender, and reorients it instead on the disciplinary techniques by which the perversions are constituted as specific pathologies. In a clever linguistic reversal, the quality ascribed to the nature of perversion—fixation—is attributed instead to the West's determined fascination with rules governing the meanings of sexuality. Here Foucault draws attention at once to the workings of power in the framing of normality and abnormality, and to the importance of the place occupied by the perversions in the history of sexuality.

The lecture John Money delivered on the occasion of the inauguration of the Society for the Scientific Study of Sexuality's "John Money Award for Significant Contributions to Sexology" on February 12, 2002, also foregrounds the study of the perversions as key to the history of sexuality. It has as its title "History, Causality and Sexology." It is, in brief, a recap of the development of the perversion/paraphilia diagnosis, which assesses the principles underlying the work of foundational nineteenth-century sexological pioneers such as Alfred Binet, Karl Ulrichs, and Richard von Krafft-Ebing—and warns against allowing the precepts from which they conducted their science to continue to influence contemporary sex research. In it, Money tells his audience that "those who cannot remember the past are condemned to repeat it."[1] The main target of his criticism of past sexology is its hand-wringing over the vexed question of whether "sexual abnormality" (including the perversions and homosexuality) is acquired or innate; he also takes the opportunity to chastise his contemporaries for being insufficiently aware of the history of their discipline. He writes: "Sexology today is, through inattention to its own historical doctrines, condemned to repeat them, as is evident in the contemporary antithesis between the biomedical versus the social constructionist model of causality in sexology, which is a reincarnation of the outmoded antithesis between nature and nurture."[2]

Money is concerned throughout his whole career with warning against false dichotomies, especially in the form of the commonplaces of nature/nurture and body/mind, as seen in the following extract from a reader's report he wrote in 1969 on an article called "Personality Characteristics of Male Transvestites: II" by P. M. Bentler and Charles Prince: "The authors outmodedly imply that cyproterone acetate works only if the symptoms are organic and not psychogenic, *but they should not align themselves in favour of a false dichotomisation, which this is, especially with reference to sex and the brain.*"[3] Yet, despite his many claims that the tendency to assume a separation of mind and body or to make a choice between nature or construction is a legacy of the past that must be overcome, we have seen in chapter 1 that Money's own relationship with biological and constructionist explanations for gender identity is complex and often contradictory, rather than systematically holistic. In what follows, I shall examine the extent to which this tendency is replicated in his theorization of paraphilia.

That Money chose to make perversion/paraphilia the subject of his milestone lecture is testament to his view of the importance of this concept and diagnosis to the history and contemporary study of sexuality, echoing Foucault's point in the epigraph, albeit from a radically different epistemological standpoint. Second, the fact that it constitutes a strongly worded and determined distancing of his own work from historical sexological and psychoanalytic views of perversion—and from contemporary work that he considered less critically aware of the past—is significant. Indeed, this is not the first time that Money asserts his difference from his forebears in these terms. A precursor of the substance of this lecture had already been rehearsed in his coauthored book with Margaret Lamacz, *Vandalized Lovemaps* (1989),[4] which is, of all Money's full-length books, the one that devotes most space to the paraphilias.[5] He begins that work with a chapter entitled "Concepts of

Paraphilia," which has as its first subheading "From Perversion to Paraphilia," promising a history of the conceptual term and its naming. Regarding Krafft-Ebing's assumption of the role of degeneration in explaining the perceived explosion of perversion in fin-de-siècle Europe,[6] Money writes: "His theory is built on an *a priori* assumption, namely, the existence of a sexual instinct, to which is added another *a priori* assumption, namely, the principle of degeneracy, to which is added, in turn, the principle of hereditary or constitutional taint."[7] He goes on: "Krafft-Ebing's theoretical formulations and their shortcomings set the agenda that his successors are still working on, more than a century later. . . . The items on the agenda . . . are the principles of hereditary, phylogenetic, neuropathological, associative, intrapsychic, and biographical determinism."[8] This sounds like a thorough diagnosis of sexology's intellectual derivation and continued preoccupations, potentially allowing Money's work to be clearly distinguished from what has gone before. Indeed, Money goes on to define what he sees as the flaw in this agenda: it treats "the paraphilias biomedically instead of criminologically."[9] The impression one gains here, in the 1989 text, is that Money is calling for a social sciences–inflected approach to paraphilia, rather than one relying on biological explanations and treatment. However, by the time he writes his 2002 lecture, he calls rather for

> A developmental theory based on longitudinal, not cross-sectional studies. Such a theory will, of necessity, be not univariate, but multivariate. The variables will be genomic status, hormonal history (prenatal and postnatal); sexual brain cell functioning; history of toxic, infectious, or traumatic exposure; infantile pairbonding; juvenile troopbonding; juvenile sexual rehearsal play; sex education; adolescent sexual history; amative history in imagery, ideation and practice and so on.[10]

It is unclear to me how these concepts differ radically from the previously denigrated biomedical/social soup of "hereditary, phylogenetic, neuropathological, associative, intrapsychic, and biographical determinism," other than in the fact that they are recast in Money-coined terminology ("pairbonding," "rehearsal play," etc.), and it is intriguing that whereas in 1989 Money appears to be pushing mainly constructionist concerns, by 2002 he foregrounds biomedical concerns, placing the twenty-first-century buzzword "genomics" as the first consideration.

In what follows, I shall question the extent to which Money is in fact able to distance his own sexological precepts from the historically inherited considerations or "agenda items" he listed in *Vandalized Lovemaps*. I shall show too that Money frequently deploys *either* "biomedical" or "developmental" explanations for paraphilia, rather than the idealized synthesis he lauds. (In fact, he has recourse, when convenient, to versions of each of the "outmoded" causal "principles" he names in the list in *Vandalized Lovemaps*, in isolation from each other, which makes his apparent blanket dismissal of them all the more puzzling.)

The aim of this chapter, then, is to see how Money positions himself at different points in his career, and in the service of different debates, with regard to the various currents in "perversion"/"paraphilia" theory that held sway at given moments over the hundred years from Krafft-Ebing to John Money. This chapter is to some extent a corrective genealogy. It focuses both on the sexological tradition in which Money is located and the parallel psychoanalytic branch of theorization and treatment with which it (and he) had an uneasy relationship. In particular, it is instructive to compare Money's work with that of his psychoanalytic contemporary Robert Stoller, as both men contributed to the American Psychiatric Association's definition of the paraphilia diagnosis in the *DSM-III-R* in 1987. I should probably add the caveat that it is, of course, the case that there are significant differences between the intellectual frameworks and scientific beliefs according to which perversions/paraphilias are conceptualized a century apart, and on different continents, and, while pointing out strategic and procedural similarities, it would be erroneous to downplay these divergences. However, John Money's work on paraphilia displays striking echoes of nineteenth-century logic, such that it would be equally misleading to accept unquestioningly the radical and complete paradigm break that Money claims for his own sexological contribution. It bears noting that where the content differs, often the logic and structure are similar, given that disciplines typically constitute themselves as such by establishing broad definitional methods and structures that may carry across temporal and ideological shifts. They provide an enduring and problematical framework for defining normality or abnormality—in this case sexual.

## NINETEENTH-CENTURY PERVERSION THEORY AND MONEY'S DEBT

The early European sexologists termed nonnormative sexual practices "perversions," a diagnostic label that persists in psychoanalytic terminology to the present day, while Anglo-American sexological and psychiatric orthodoxy adopted "paraphilia" (literally "beside love" or "beyond love"), after a suggestion by Wilhelm Stekel in 1908. "Paraphilia" is thought to have been used for the first time in the English language in 1934 by psychiatrist Benjamin Karpman.[11] The nominal intention behind introducing the term "paraphilia" was both to free psychiatric terminology from its proximity to psychoanalytic concepts and to reject the originally religious implications of "perversion" as a moral "turning aside" from the path of righteousness. In John Money's words, paraphilia is "a biomedically impartial synonym for the morally judgmental term 'perversion.'"[12]

Arnold Davidson has shown that the introduction of the concept of "perversion" with reference to sexuality in nineteenth-century medical circles was predicated on the dominant understanding of the human sexual instinct as identical with the drive for reproduction, ensuring the preservation of the species.[13] This understanding of the sexual instinct as reducible to the drive for reproduction can be found both in the work of French doctor Paul Moreau de Tours and in Krafft-Ebing's *Psychopathia Sexualis* (1886), considered the first psychiatric work on perversion, which itself was heavily influenced by Moreau de Tours's *Des aberrations du sens génésique* (1877).[14] According to this narrow and somewhat utilitarian understanding of the functioning of sexual instinct or *instinct génésique* (always pulling toward the opposite sex, because always seeking propagation of the species), everything except penetrative heterosexual intercourse would logically come under suspicion as abnormal or *contra nature*.

Krafft-Ebing develops, from the basic principles set out by Moreau de Tours, four subclassifications of sexual "anomaly": (1) sexual anesthesia, in which the desire for sexual activity is entirely lacking (e.g., frigidity); (2) hyperesthesia, in which the intensity of sexual desire is excessive (e.g., nymphomania); (3) paradoxia, in which the sexual instinct is present at an inappropriate (nonreproductively viable) age, namely, in infancy or very old age; and finally (4) paresthesia, in which the sexual instinct does not lead to a reproductive act, but instead turns aside (is "perverted") in the direction of some other act. Homosexual acts and perversions/paraphilias would both come under this fourth category, given the narrow understanding of the "proper" nature of sexuality as always already reproductive.[15]

The result of employing these rigid definitional parameters around normality was that a self-fulfilling prophecy was constituted. It becomes an apparently observable fact that abnormality, in taking so many forms, in being everything that is not the one thing that is "correct," appears to be everywhere, endlessly proliferating, and out of control. Thus, the hypothesis about human sociosexual life that the sexologists most feared appears to be true precisely because of the premises upon which they construct their logic.

The *Psychopathia Sexualis* is, as the name suggests, a textbook about sexual malady, not sexual health. However, in excluding so much from the potential remit of "health," it gives the perhaps unwitting impression that sexual pathology is more pervasive than sexual "normality." This is but one legacy that Krafft-Ebing leaves to subsequent sexual science. This long-lasting feature can be understood along the lines of Georges Canguilhem's analysis (1966) of the concept of "normality" in early forms of modern medical and social science.[16] According to Canguilhem, "normality" was supposed to describe a hypothetical, statistically "average man," what Vernon Rosario has called "a demographic construct rather than any particular person."[17] However, this construct soon came to be used not as a descriptive fiction but as a *pre*scriptive one, with moral weight attached to it. Normality, in short, moved from meaning the most commonly occurring, or the statistically average, to mean "the ideal," with all the rarity—if not material impossibility—suggested by the Platonic overtones of that term.

In this respect, John Money's textbooks of paraphilia, *Lovemaps* and *Vandalized Lovemaps*, proceed in generically and strategically similar ways, and produce similar effects, to *Psychopathia Sexualis*. One of the very few points of praise that Money has for Krafft-Ebing as a sexological pioneer is the fact that he collected "more extensive clinical data on the paraphilias than had ever before been recorded."[18] This comes in the form of an exhaustive series of case studies, and the coining of names for individual perversion types. Money's work on paraphilia too is characterized by a passion for providing technical names (often ugly Greco-Latinate portmanteau words) for scores of sexual practices, producing an overwhelming catalog of varieties and subsets of paraphilia. Where Krafft-Ebing famously gives a name to "sadism," for example, Money provides us with such neologisms as "apotemnophilia" and "autassassinophilia." Money thus offers a twentieth-century taxonomy of the paraphilias that resembles Krafft-Ebing's taxonomic nosography of the perversions from exactly a century earlier. When consulted about how the paraphilia diagnosis could be improved for the third edition revised of the *DSM*, Money sends a list of the paraphilias he has named for inclusion, insisting on the need for comprehensive taxonomy, a wholly nineteenth-century gesture.[19] And, where Krafft-Ebing separated errant sexuality into the four subclasses of "anomaly," of which one becomes perversion/paraphilia, Money takes this further, dividing the paraphilias *themselves* into overarching categories and naming "six grand paraphilic stratagems" ("sacrificial/expiatory"; "marauding/predatory"; "mercantile/venal," "fetishistic/talismanic"; "stigmatic/eligibilic"; "solicitational/allurative"), into which the individualized practices and fantasies are then fitted.[20] Given these textual features of both corpuses, it is easy to see that what Foucault called "a steady proliferation of discourses concerned with sex"[21] when describing foundational European sexology is equally visible in the twentieth-century work of the Johns Hopkins paraphilia theorist.

While in aspects such as their shared "taxonomania," it is quite easy to see what Money has in common with Krafft-Ebing, on other points there is more ambiguity. One such case is that of the extent to which social pressures and norms are admitted as having a strong influence over the development of individual "normal" or "abnormal" sexuality. We have seen that the nineteenth-century sexologists proceeded from the assumption that sexuality was naturally identical with the biological instinct for reproduction, suggesting a dependence on (what Money calls) "phylogenetic determinism." Krafft-Ebing writes: "the propagation of the human species is not committed to accident or the caprices of the individual, but made secure in a natural instinct which, with an all-conquering force and might, demands fulfilment."[22] Yet, it would be too easy and fallacious to conclude that, in the nineteenth century, "natural instinct" is given absolute sovereignty as the only explicatory mechanism for how human sexuality works. In the *Psychopathia Sexualis*, there is the suggestion that social conditioning plays a very important part in the achievement of sexual normality. Indeed, the work's introduction constitutes a programmatic account of how social institutions and behavior shape "healthy" sexuality.[23]

> In coarse, sensual love, in the lustful impulse to satisfy the natural instinct, man stands on a level with the animal; but it is given to him to raise himself to a height where this natural instinct no longer makes him a slave: higher, nobler feelings are awakened which, notwithstanding their sensual origin, expand into a world of beauty, sublimity, and morality. On this height, man overcomes his natural instinct.[24]

The "height" of civilized feeling to which Krafft-Ebing refers can be found, he goes on to suggest, via the religious and social institution of marriage, which is held up as the only appropriate domain for sexual expression. In this, Krafft-Ebing is very clearly paying lip service to religious discourses, perhaps as an attempt to overcome the suspicion of immorality that accrued to early sexual science, owing to the nature of its object of investigation.[25] Along with the necessity of marriage for ensuring sexual health, Krafft-Ebing counsels censorship of pornographic material, leading him to write of the licentious works of the Marquis de Sade that "fortunately it is difficult to-day to obtain copies."[26] Sexuality here is seen as an ethical project, then, an achievement of civilization, which in and of itself stands at odds with the other idealizing discourse whereby the most "normal" sexuality would approximate a "natural" sexuality.

At first glance, it would seem that Money's work could not be more different from Krafft-Ebing's in this respect. Money writes much of healthy sexual variation, of experimentation, and of pornography as positive, liberating forces. On the question of contagion or corruption by exposure to explicit material, which was the counterpart to degeneration in nineteenth-century logic (and that led Krafft-Ebing to render particularly licentious details in Latin rather than German, such that the lower orders may not be corrupted), Money writes: "[the idea of contagion] is the phobia that has paralyzed society into a state of obsessional indecision regarding the sort of material that is acceptable or safe, and the kind that is not."[27] In a spirited defense of anticensorship, with specific regard to pornography, he goes on:

> A rational way for society to deal with pornography is the antithesis of phobia and criminalization, and also the antithesis of unsupervised laissez faire. . . . It would require a new generation of society emancipated from the contagion and degeneracy theory of sex and eroticism.[28]

Here, then, we are led to understand that Money's enlightened social agenda is in contradistinction to Krafft-Ebing's "unenlightened" attitude to censorship as seen in his comments about Sade. However, the discursive and didactic strategies of the two are largely the same; only the message is different: where Krafft-Ebing counsels making a good wholesome marriage and avoiding pornography as the path to "health," Money counsels "Sexual Revolution"-inspired erotic experimentation and the consumption of pornography as "healthy." In both cases, a male authority figure is prescribing the pursuit of historically located, ideologically fashionable practices, under the guise of science, with no explicit analysis of the power dynamics at play in the production of these cultural ideals. Moreover, while Money's propornography, sex-positive politics allow him to mark a clear distance from his historical forefathers in terms of ideological agenda, he does not actually get rid of the idea of corruption (negative influence on healthy sexual development) altogether. Rather, he cites moralistic interdictions on sexual representation, and the very perpetuation of the discourse of contagion or corruption itself, as causes of sexuality gone awry—or (somewhat contradictorily)—of sexuality corrupted.

It is important at this point to ask what exactly it is that "goes awry" or "gets corrupted" in the case of paraphilia for Money, if not the "natural instinct to procreate" or "*instinct génésique*." Central to the apparent, ambivalent constructionist thrust in Money's sexology is his concept of the "lovemap," which provides the title of his two well-known works on paraphilia.[29] The lovemap is, Money states, like "a native language," in that it develops several years after birth. It is "a developmental representation or template in your mind/brain." It is worth noting the conceptual ambiguity kept in play between the psyche and the brain, which Money restates elsewhere as "Lovemaps are an example of both cognitive mapping and brain mapping."[30] The lovemap "depicts your idealized lover and what, as a pair, you do together in the idealized romantic, erotic and sexual relationship."[31] Superficially at least, then, a move is visible from an essentialist understanding of the natural instinct for procreation that can be perverted by inherited aberration or corruption, toward a more developmental model of sexuality analogized to linguistic acquisition. Paraphilia occurs, according to Money, when the lovemap is inhibited from forming normally or is "vandalized." Vandalization may come about as a result of abuse, of a traumatic experience that then gets eroticized in the service of preserving sexual feeling, or simply owing to inadequate education about sex and lack of appropriate "rehearsal play" with other children.

Money defines paraphilia as

> A condition occurring in men and women of being compulsively responsive to and obligatively dependent upon an unusual or personally or socially unacceptable stimulus, perceived or in the imagery of fantasy, for optimal initiation and maintenance of erotosexual arousal and the facilitation or attainment of orgasm. Paraphilic imagery may be replayed in fantasy during solo masturbation or intercourse with a partner. In legal terminology, a paraphilia is a perversion or deviancy; and in the vernacular it is kinky or bizarre sex.[32]

In *Lovemaps*, Money explains in detail his hypothetical theory of the genesis of paraphilia as deformed or corrupted lovemap development, which differs (in content but not structure) from the nineteenth-century understanding of a model of *procreative instinct* gone awry. Both models presuppose a correct and superior, if not in the case of Money *originary or natural*, sexuality, to which perversion/paraphilia is the improper counterpart.

We have seen that for the nineteenth-century sexologists, one of the precepts of the notion of perversion as the effect of inherited degeneration was that sexual perversion would likely occur in those subjects with other types of physical or moral so-called abnormality, passed on from parents or forebears. This obtained both in Krafft-Ebing's nosography, in which both perverse patients and their families would be examined for nervous, physiological and developmental disorders and illnesses, and especially in Lombroso's criminal anthropology, in which the body would be measured to reveal the concomitant inborn moral, sexual, criminal taint. Given the precepts on which the theory of "lovemap" seem to rest, we might assume that such logic could have no place in Money's paraphilia theory. Yet, in fact, Money does not move very far away from this. In *Vandalized Lovemaps*, alongside Money's nominally environmental, constructionist theory that paraphilias are acquired as a result of traumatization or vandalization of a child's lovemap (as the book's title suggests), we see an entirely different and contradictory discourse about the determining link between physiology and sexuality. In Money's system, the idea of the sexually "normal" body goes hand in hand with "normal" desires and behaviors. Conversely, at points in Money's work, paraphilia seems to be predicated upon chromosomal and physiological "abnormalities," conditions of intersex, or neurological damage.

Not a single case study discussed in *Vandalized Lovemaps* is "simply" a case study of a "paraphiliac." A glance at the contents page reveals that the cases discussed include "Sadomasochism in a Male with Congenital Micropenis"; "Pedophilia in a male with a history of hypothyroidism"; and "Bondage and discipline in a female with congenital vaginal atresia." Similarly, on the link between paraphilia and neurological abnormality, Money writes: "in the clinic, it is not uncommon for a patient to have a dual diagnosis of epilepsy and paraphilia, and to have a history of epileptic attacks separate from paraphilic fugue states. Likewise, it is not uncommon for a paraphilic diagnosis to coexist with a history of traumatic head injury or with a history of manifest neurological dysfunction."[33] The reference to the proximity between paraphilic fugue states and epilepsy resembles nothing so much as Lombroso's studies of so-called inborn criminals, in which epilepsy was considered to be a key symptom of degeneration and often diagnosed in criminals and perverts.[34] Moreover, Money misses the rather obvious fact that the people seen in a neurological clinic would tend, by necessity, to be those with neurological damage/disorders, some of whom might *also* have a nonnormative sexuality. The idea that there may also be large swathes of the population *not* seeking help for neurological, chromosomal, or physiological conditions who are behaving sexually in ways that Money would term "paraphilic" is simply not countenanced. For all his tokenistic rejection of inherited degenerate traits when named as such by Krafft-Ebing or Lombroso, Money repeatedly links paraphilia—the sick counterpart to healthy sexual desire—to other(ed), nonnormative bodies and states of congenital pathology throughout *Vandalized Lovemaps*.

It is notable that Sigmund Freud was one of the few turn-of-the-century sexual scientists to question the widespread assumption of the link between degeneration and perversion. He writes in 1905:

> It is natural that medical men, who first studied perversions in outstanding examples and under special conditions, should have been inclined to regard them as indications of degeneracy or disease. Nevertheless, it is even easier to dispose of that view in this case than in that of inversion.[35]

And, regarding the idea that sexual perversion will be necessarily accompanied by other systemic physiological, neurological, or psychiatric illnesses, Freud is equally keen to debunk contemporaneous assumptions. Writing of the state of the mental health of even the most "extreme" perverts, Freud asserts: "Certain of [the perversions] are so far removed from the normal in their content that we cannot avoid pronouncing them 'pathological.'" (The examples he gives are necrophilia and licking excrement.) Yet, he goes on:

> Even in such cases we should not be too ready to assume that people who act in this way will necessarily turn out to be insane or subject to grave abnormalities of other kinds. . . . People whose behaviour is otherwise normal can, under the domination of the most unruly of all the perversions, put themselves in the category of sick persons in the single sphere of sexual life.[36]

Despite his searing criticism of psychoanalysis's repressive hypothesis, Foucault commented approvingly in 1976 that psychoanalysis, unlike sexology, "rigorously opposed the political and institutional effects of the perversion-heredity-degeneration system."[37] John Money, on the other hand, while refuting the specifics of degeneration as an "outmoded" ideological discourse, refuses to acknowledge the radicality of Freud's rejection of the assumption that sexual abnormality must be accompanied by other forms of abnormality. In his 2002 lecture, he writes: "Although Freud did not follow Krafft-Ebing's example and postulate hereditary weakness or taintedness as a predisposition toward sexual perversion . . . logistically he could have done so."[38] This is a contrary misreading, since while Freud could have *logistically* accepted degeneration, *politically and ideologically*, as Foucault points out, he would not have done so, since the very notion of degeneration condemned the individual to a predetermined sickness. This would have been a politically dangerous hypothesis to entertain, since the eugenic experiments of the Holocaust had their origins in the belief in inherited degenerate taint. Freud, by contrast, and for all his blind spots and shortcomings with regard to, for example, female sexuality, seemed to understand, like Foucault, that scientific theories and hypotheses *are political*. Freud's model also allows for the possibility that psychic states are motile and open to change. Money's unwillingness to appreciate the import of the gesture Freud makes in condemning the ubiquitous fashion for degeneration and the linking of perversion with physiological and psychiatric illness suggests a refusal to accept fully the ideological character of medical theory. It also suggests that Money is hankering, *contre-cœur*, after a biologically determined origin for paraphilia, despite his own professed investment in the conversion of "vandalized lovemaps" to normal ones, which clearly borrows from the methodology of the "talking cure."

## "INTRAPSYCHIC DETERMINISM" AND MONEY'S CRITIQUE OF FREUDIAN PSYCHOANALYSIS

John Money maintains in *Vandalized Lovemaps* that his principal reason for dismissing Freud's psychoanalytic theory of perversion is that it rests on what he calls "endopsychic" or "intrapsychic" determinism. That is, he claims that psychoanalytic theory looks only at the interior psychical development of the individual, presupposing a set of developmental stages, and a range of possible (neurotic/perverse) outcomes. He criticizes psychoanalysis for excluding environmental factors, which Money terms "extrinsic" (such as traumatic interference or abuse) and congenital factors (such as neuropathology, but also—and these are direct examples Money gives—"degeneracy" and "hereditary taint"). Money claims, moreover, that "the hazard of intrapsychic determinism is that . . . it is too readily converted into a dogma. A dogma is validated by the number of its converts and, conversely, of its victims, rather than by the pragmatics of its empirical productivity."[39] The refusal to see the capacity for *any* theory, including Money's own, produced within an authority discipline, to become a dogma with slavish adherents is regrettable. (A generous interpretation of the many apparent logical contradictions found in Money's œuvre is that they might be deliberate features, intended to prevent the hypotheses and theories from solidifying into systematization and thereby into dogma. But I suspect this may indeed be too generous.)

Of Freud himself, Money writes in the 2002 SSSS lecture:

> His self-appointed task was to formulate an exclusively endopsychic explanation of perversion and sexuality in general—not in terms of sexual practices only, but also in imagery and ideation, conscious and unconscious.... Freud did not find a satisfactory answer to the question of who would be predisposed to develop either a perversion, or a neurosis, or neither. There probably is no answer within the dynamics of an exclusively endopsychic theory. In other words, endopsychic theory is a universe of discourse unto itself.[40]

Money's stated investment in the importance of being able to predict inherent predetermination toward perversion, and his critique of Freud for failing to take account of this, seem to hark back entirely to the worldview of Krafft-Ebing and his contemporaries, who Money elsewhere disavows (as seen in his citing of "degeneracy" and "hereditary taint" as considerations that are deleteriously missing from psychoanalysis, which is simply odd).

Foucault's criticism of nineteenth-century sexology's "specification of individuals," which resulted in defining the "perverse personage," is in the service of wishing to question the problematic idea that a subject's sexual desires and acts define their entire identity. Money, by contrast, seeks to know "who would be predisposed to develop ... a perversion," suggesting that perverts are born rather than made. In propounding this deterministic idea, with eminently nineteenth-century roots, he risks undermining the importance he has elsewhere accorded to the vandalization of lovemaps, or the Christian repression of sexual expression (itself a rather psychoanalytic idea), as causes of sexual paraphilia. Surprisingly, perhaps, for a turn-of-the-century thinker, Sigmund Freud may in some ways be read as more "progressive" than Money in his formulation of perversion, since, as we have seen, he sought to repudiate predetermination to perversion in the form of degeneration or inherited abnormality—a move Money was reluctant to make, even as he condemns the outmoded discourse of degeneration as such.

However, following a Foucauldian logic, we might propose that a given discipline's treatment regimen necessarily produces the etiology of the illness, rather than the other way round. In this sense, in the specific case of perversion/paraphilia, as Freud's claim is for the curative benefit of a purely talking cure, so it behooves psychoanalysis to insist on the "endopsychic" as the location of the perversion, since this is where the cure too will lie. Whereas, since Money proposes a combined program of talking therapy (to "straighten out the lovemap"), combined with antiandrogen drug therapy, this suggests the absolute pragmatic necessity of positing dually hormonal and psychological causes for paraphilia.

Moreover, just as Money's disavowed debt to nineteenth-century sexological theory has been traced above, so it is possible to uncover in Money's corpus elements of psychoanalytic logic and language that are not reconciled with the rest of his system, and that stand in contradiction with his claims about the particular failings of psychoanalysis. Perhaps this is most explicitly visible in Money's earliest works such as *The Psychologic Study of Man* of 1957, which uses the term "ego," most usually associated with Freudian psychoanalysis, as a central psychological principle. The book features chapter headings such as "Ego Function of Spectatorship" and "Ego Pathologies." Money first criticizes Freud for devising a system in which "mind and body were as separate and distinct as they had ever been."[41] He then states that the term "ego" needs to be understood in such a way that it is not identical with the psychoanalytic concept, but rather that it may serve the same function as his own coining "bodymind." The purpose of coining both "bodymind" and Money's refashioned "ego" is to "ensure that outmoded connotations of separateness between mind and body do not sneak in through the back door."[42] There is something rather perverse, however, in the use of a Freudian term *to ensure the avoidance of something Money (mis)-ascribes to Freud.* All the more strangely for a former lecturer in psychology, Money demonstrates either ignorance or a willful misrepresentation of the psychoanalyst's position, in choosing not to discuss Freud's explicit claim that the mind and the body are very definitely not separate, since the ego must be understood as "first and foremost a body-ego."[43] Moreover, in choosing to ignore Freud's well-documented interest in, and (however inconclusive) speculations on, the interaction of biology and the psyche, Money is guilty with regard to psychoanalysis of characterizing it inaccurately as resting on the false dichotomization he so often critiques.[44]

To make more explicit the argument that Money often apes psychoanalytic logic while repudiating the name of psychoanalysis, we might turn to a paper from 1987, "Masochism: On the Childhood Origin of Paraphilia, Opponent-Process Theory, and Antiandrogen Treatment."[45] In this paper, Money describes a letter from an Indian correspondent that tells how, as children at school, he and his classmates were regularly caned by the headmaster, sometimes with the master's attractive wife looking on. The letter writer proceeds to recount how some of his peers went on to achieve orgasm as a result of canings, and to develop a dependence upon this type of stimulation, often seeking out prostitutes to fulfill this need. Others turned to self-flagellation rather than employing the services of a paid partner.

Money sees the latter resort as particularly "masculine" or "macho" in its lack of reliance on another person. He also sees it as falling within a tradition of religious zeal, in which the mortification of the flesh is a form of spiritual exercise.[46] As is all too common in Money's work, we see evidence of contradiction and partiality here since, in this case, environmental and cultural factors are seen as *entirely responsible* for the genesis of the paraphilia in question, without regard to the physiological and hormonal factors on which he elsewhere insists (principally in *Vandalized Lovemaps*). Indeed, this version of Money-esque paraphilia is wholly developmental and, I would venture, "intrapsychic" since only some of the schoolfellows become erotically inclined toward flagellation, suggesting that this is one possible psychosexual path that might be taken. So Money's case study is at least as guilty as Freud's writing on perversion of the charge he levels at Freud of failing to account for why only *some* individuals who are exposed to certain environmental factors become paraphiliacs.

It is particularly noteworthy that Money's theoretical paper on the development of masochism bears numerous similarities to Freud's "A Child Is Being Beaten" (1919), which undertakes imaginative speculation about the mobilization of positions of rivalry, identification, and desire at play within phantasy when a child is faced with the spectacle of a classmate or sibling receiving corporal punishment and then eroticizes that act.[47] Money's critique of psychoanalysis for intrapsychic determinism rests on the idea that contingent environmental factors are not taken into account when looking at the etiology of a perversion. But this is blatantly false. In Freud's classic account of his prototypical perversion, fetishism, in 1927, had the child not contingently glanced upon a nose that took his fancy, the particular perverse outcome would not have been achieved in adulthood.[48] Freud ingeniously argues that the fetish is linguistic in formation. A "glance at the nose," first stolen in the child's British nursery and overvalued, became the fetish for a shiny nose (*Glanz auf der Nase*) of the adult patient who had long since been removed to Germany. We might note the parallels between Freud's notion of the fetish as linguistically determined and Money's idea that the lovemap itself develops "like a native language." Similarly, had the analysands described in "A Child Is Being Beaten" not witnessed other children being punished at school, much as Money's correspondent had, the masochistic desire would presumably not have taken the form it did. In fact, when considering the part played by extrinsic environmental factors in the etiology of perversion, Freud's and Money's essays on the formation of masochistic fantasy seem to accord it pretty much equal importance.

To make one further and final point about Money's crypto-psychoanalytic essay on the formation of masochism, he takes from the letter written by his Indian correspondent corroboration of his pet theory with regard to the dynamic of paraphilia, namely, that it is "the reconciliation of opposites. The negative becomes positive. Tragedy becomes triumph. Aversion becomes addiction."[49] Money goes on: "The price of this reconciliation is the severance of lust from pair-bonded love. It is typical of paraphilia that the lust partner and the love partner are not the same person. That is why it is more accurate to say that paraphilias are love disorders rather than lust disorders."[50] In this formulation, Money's theory of paraphilia as a "conversion" of trauma into orgasm retains a psychoanalytic flavor, but as well as resembling the mechanism described in "A Child Is Being Beaten," it even more closely approximates Robert J. Stoller's concept of perversion as the conversion of sexuality into hostility. I will expand upon this contention in the next section by comparing Money's paraphilia with Stoller's perversion.[51]

## "THE EROTIC FORM OF HATRED": MONEY AND STOLLER ON PERVERSION/PARAPHILIA

Money's contemporary, Stoller, fellow consultant on the *DSM-III-R*, and equally influential contributor to the psychiatric understanding of "paraphilia" in the period 1970–90, articulates a theory of perversion in the psychoanalytic tradition that resonates with Money's notion that "paraphilias are love disorders rather than lust disorders" and that defines the person "suffering from" a perversion as a particular sort of psychological personage whose condition prevents intimacy, love, and altruism.

Stoller's most famous work, *Perversion: The Erotic Form of Hatred* (1975), posits that the defining feature of "perversion," by which "one can recognize [it] when it appears," is "hostility."[52] Stoller explains: "*Perversion*, the erotic form of hatred, is a fantasy, usually acted out but occasionally restricted to a daydream. . . . It is a habitual, preferred aberration necessary for one's full satisfaction, primarily motivated by hostility. . . . The hostility in perversion form is a fantasy of revenge hidden in the actions that make up the perversion and serves to *convert childhood trauma to adult triumph*"[53] and "in the perverse act the past is rubbed out. This time trauma is turned into pleasure, orgasm, victory."[54] We can see a similarity here between Money's formulation discussed above—"The negative becomes positive. Tragedy becomes triumph. Aversion becomes addiction"—and Stoller's pronouncement. Given that Stoller's canonical text was first published in 1975, while Money's major works on paraphilia date from about ten years later, it is likely that Money has borrowed Stoller's logic, which, notwithstanding its psychoanalytic framing, fulfills Money's apparent wish to maintain a (not well defined) dual causal relationship between brain/hormonal abnormality, on the one hand, and childhood trauma, on the other, as determining the outcome of perversion/paraphilia.

Stoller's *Perversion* is replete with language that, to the critical contemporary reader, appears freighted with ideology that we might term both homophobic and "kinkphobic." A prime example is his formulation "such obvious perversions as rape, exhibitionism, sadism, or homosexuality."[55] The blatant prejudice underlying the strategy of aligning homosexuality unquestioningly with rape—a nonconsensual sexual attack—is regrettable to say the least. Similarly, pro-BDSM activists such as medical doctor Charles Moser, himself a colleague of Money and Stoller, would argue that a preference for sadism, if carried out with a consensual, masochistic partner, should not be understood as necessarily pathological—much less criminal, like rape.[56]

Money, by contrast, is keen to assert that homosexuality is *not* a paraphilia (aligning him more closely, in this instance, with Freud than the later psychoanalyst Stoller). For example, he writes in 1998 that

> The homosexual person, like the bisexual or heterosexual, may be either normophilic or paraphilic. Paraphilia is independent of homosexuality or heterosexuality. If a paraphilia afflicts a heterosexual person, it is the paraphilia that needs treatment, not the heterosexuality. Likewise, it is the paraphilia and not the homosexuality that needs treatment when a homosexual person is paraphilically afflicted."[57]

Money seems here to be determinedly dislocating the sex/gender-orientation of a person's sexual attraction from the sexual practices they (are compelled to) perform. Yet, in the abstract of a coauthored article on apotemnophilia (erotic self-demand amputation), it is stated: "the apotemnophiliac obsession . . . may be conceptually related to, though it is not identical with, transsexualism, bisexuality, Münchausen syndrome, and masochism."[58] This is a strange formulation that resembles Stoller's grouping together of rape, sadism, and homosexuality. It *conceptually* aligns an identity disorder (in psychiatric ideology), a sexual orientation, a mental disorder, and a paraphilia as characterized by "obsession." ("Obsession" is a close synonym for "compulsion" or "obligative condition"—terms Money repeatedly uses to describe paraphilia's "frozen countenance," to return to Foucault's critical vocabulary.) Given that Money himself was fairly openly bisexual (and described as such by Richard Green in his commemorative article on Money, Stoller, and Harry Benjamin[59]), this linking of pathological, obsessive paraphilia with what is elsewhere assumed to be a harmless orientation is a lapse—and typical of Money's propensity to inconsistency.

Stoller seems to operate more squarely in the fairly homophobic tradition of much post-Freudian psychoanalysis.[60] Yet he works hard to make clear that while considering homosexuality a perversion, and heterosexuality the preferable, healthy, and "mature" resort, he understands that heterosexuality is not "natural" or inevitable. On the one hand, then, he calls perversion nothing but "blighted heterosexuality";[61] while on the other, he avows that "heterosexuality is an acquisition; we cannot brush the issue aside by saying that heterosexuality is preordained, necessary for the survival of the species and therefore biologically guaranteed."[62] This bet-hedging equivocation between an acknowledgment of the workings of contingent cultural normativity in prioritizing heterosexuality and a contrary assertion of its superiority is typical of sexological accounts from the nineteenth century to the present day. Stoller and Money share this rhetorical technique of wishing to keep in simultaneous play ideologically incompatible ideas dressed as objective science.

By starting from the point of view that "normal" sexuality is (implicitly for Money; explicitly for Stoller) heterosexual and reproductive, twentieth-century sexological interventions in paraphilia ape the ideology dressed as science of the nineteenth century. Moreover in their assumption that medicine has the ethical right to impose "normality" on the sexual subject, both projects seem disturbing from the perspective of a post-Foucauldian or queer reader—for all that they use a language of social liberalism to distinguish themselves from those authority voices who would dismiss perversion as "sin." Religion, indeed, is the prime target for both Money and Stoller. Stoller describes the idea that perversion is merely sinful sexuality as "the product of a Judeo-Christian heritage, fortified in each different generation and place by local conditions in the service of bigots. . . . When one changes the beliefs of society, the sense of sin will dissipate";[63] while Money opines in 1988 in a letter to a colleague with whom he has a correspondence on pornography: "I have come to the conclusion that everything about pornography is so religious, moral, political, legislative, and judicial, and so epistemologically chaotic, that it is beyond science. It's like trying to argue against the death sentence while the prisoner is already in the death chair."[64] Religious authority thus serves as an irrational, outmoded benchmark against which to assert the progressive, rational, modern neutrality of science. As in the case of Money's expressed desire to obliterate American religiosity, which he claims in *Vandalized Lovemaps* is "on another crusade . . . against the heresies of the sexual revolution,"[65] Stoller's book has an apparently radical, social reformist conclusion. Stoller argues that the nuclear family itself produces perversion, and that doing away with the family would decrease the frequency of the condition: "Not knowing what will come if the family disappears, we cannot know how human sexuality will, in adapting, be modified. My guess is that if all goes well for our race, perversion will die down and variance increase. Perhaps some day perversion will not be necessary."[66] Liberal ideas about social and sexual reform coexist uncomfortably, then, with very normative ideas about gender identity and perversion in both Stoller's work and the work of John Money (a theme I will discuss with regard to Money in more detail in chapter 6).

Like Money, Stoller seems keen to abstract sexual abnormality from what Money terms Freud's "intraspsychic" concerns. Stoller writes that "aberrant sexual behaviour . . . is ubiquitous in man, and is the product of brain and hormonal factors that can function independently of anything we might call psyche."[67] Stoller actually separates "sexual aberration" found in people with "hermaphroditic identity" and those with hormonal and chromosomal abnormalities (women with androgen insensitivity syndrome; people with Klinefelter's syndrome) from "perversion" that is the result of childhood trauma (what Money will call "lovemap vandalization") in the hormonally and chromosomally "normal" person (a convenient fictional character). Those whose chromosomes, hormones, or genitalia do not fit the normative binary template are considered instead to be likely to suffer from aberrations of both sexual desire *and* gender identity. This is a problematic logic, as it first assumes that the psychically "normal" person is the person who fits a perfect binary idea of biological maleness or femaleness—and of corresponding psychical masculinity or femininity. To become a pervert, such a (potentially) normal person would need to *be perverted* in childhood. The idea that gender identity is a stable category with biological origins is necessary to Stoller's theory, since a mainstay of his argument is that "perversion arises as a way of coping with threats to one's gender identity."[68] Hence, perversion is redefined by Stoller away from the classic Freudian model of disavowal of the difference between the sexes (denoted by "mother's castration") and is seen instead as a means of protecting the sovereignty of sexual difference, while Money's paraphilia works analogously to protect a person's capacity for *sexuality itself* from the threat of annihilation it allegedly faces as a result of vandalization.[69]

## CONCLUSION

For those who associate John Money primarily with the social constructionist turn in sexual medicine in the 1970s, the prediction with which he chooses to close his 2002 SSSS lecture on the history of perversion theory is perhaps rather surprising:

> The cutting edge of research will be in the animal lab rather than in human investigation. . . . Social constructionism will be sidelined in social science and the humanities where it will be used to explain shifts in sexological ideology, rather than the development of individual sexuality.[70]

This statement is both clairvoyant and ideologically interesting in terms of what it reveals about Money's attitude to his own legacy. First, Money is accurate in his prediction that sexual medicine in the twenty-first century is much more oriented toward biological research, especially in the dual form of neurology and genetics/genomics, than toward the insights of social psychology. Second, it is striking that Money implicitly endorses this turn in what may be seen as an attempt to distance his contribution to sexuality studies and perversion theory from the taint of social constructionism—the very current of intellectual endeavor for which he became (in)famous. Perhaps he develops this attitude in his very last writings, just prior to his physical decline and death, because of the very public fallout of his most significant experiment with the implications of constructionist theory: the case of David Reimer.

Yet, as we have seen throughout this chapter, Money's theory of paraphilia has often prioritized biomedical explanations for paraphilia, while also offering a parallel (but not integrated) theory of its developmental etiology in the form of "lovemap vandalization." While evidently borrowing from Robert Stoller's formulation of perversion in numerous respects, Money does not separate cases of "aberrant sexual behavior," which are thought to occur as a result of nonnormative hormonal, chromosomal, or genital development, from "perversion" proper, as Stoller does. Rather, he superimposes paraphilia *onto* a range of intersex conditions and neurological illnesses in *Vandalized Love-*

*maps* in a gesture of having it both ways that recalls irresistibly Krafft-Ebing's simultaneous appeal to inherited degeneration *and* contagious corruption, despite the logical incompatibility between the two models.

Money repeatedly stated that he rejects *both* postulations of nineteenth-century perversion theory—inheritance and contagion—as falsehoods. First, he states, the notion of an inherited predisposition to paraphilia does not account for the very different forms and contents that paraphilia takes. Second, an understanding of paraphilia as corruption or contagion fails to account for the fact that a nonparaphilic person watching a specific kind of paraphilic pornography will not be turned on, but rather bored, disgusted, or merely intellectually curious. If paraphilia were contagious, Money argues, doctors, judges, and those who censor violent and sexual films would "catch" it. Yet, in a range of ways, it is clear that the work of John Money on paraphilia does not constitute the major paradigm break that Money claimed. Rather, it fits neatly within the tradition established in the nineteenth century whereby sexuality is constructed using the fashionable (dichotomous) discourses of the day: degeneration versus corruption/contagion for Krafft-Ebing, which becomes physiology/endocrinology versus cultural vandalization for Money. Money's constant move between favoring first environmental, then biologistic causes for paraphilia effectively perpetuates the very charge he levels at other systems, such as psychoanalysis, to the effect that they maintain a body-mind split. Indeed, Money seems to move strategically, depending on the argument he wishes to make at a given moment, between claims for contingent corruption (vandalization) as the cause of paraphilia and counterclaims that rely on the assumption of coterminous underlying and predeterministic abnormalities or pathologies, without being able to integrate them.

I want to conclude this chapter with a specific illustration of Money's tendency to vacillate between dichotomous discourses by examining an article in which he is cited, and in which the substance of his claims casts interesting light on his attitudes to the social power relations within which debates about paraphilia play out. The article by Bronwen Reid, written largely from a feminist constructionist viewpoint, was published in the *New Zealand Times* in 1983.[71] It concerns motivation for rape, and in it Money was asked to comment on the usefulness of antiandrogen treatment for rapists, posing the question of whether rape is a social phenomenon, motivated by the desire for power, or the involuntary resort of a few mentally ill men. The article tells us: "Professor Money . . . discounted the view that rape is about power not sex. [He] said rape was a specific medical syndrome. He believes 90 per cent of rapists genuinely 'can't help themselves.'" (Money has coined a term for this "specific medical syndrome." *Raptophilia* is the name he gives to paraphilic rape.) Further, Money is quoted as stating, "They don't know why they do it, they hate themselves afterward and you can't help these men till you give them a rest from the sexual drives."

After making these statements about inherent biological drives gone awry in those predisposed to paraphilia, Money turns to environment and vandalization. The article comments that "after reconstructing rapists' histories, Professor Money concluded they were raised in an atmosphere of sexual taboo 'which is just as strong in New Zealand as it is in Baltimore.'" The article tells us that Money claimed that rapists were likely to have been boys whose healthy sexual rehearsal play between the ages of five and eight had been punished, such that they grew up believing that sexuality was bad. Money is quoted as saying, "Rapists were never able to believe that sex could be had under normal caring circumstances. It could only be had if it was wicked and naughty and never with a virgin. The woman had to be kicking, fighting, biting and screaming and, more importantly, terrified."

When the author of the article points out that this sounds similar to the feminist argument that rape is about the misogynistic exercise of masculine (social) power, rather than an overwhelming paraphilic drive, Money counters by borrowing from his earlier logic of biological determinism: "There is a hormonal dysfunction. Once you can control that you can then begin to work through the fantasies that the rapist has." It is almost as if Money resorts to the cultural capital that biological scientific arguments have at the very moment when he is asked to consider the gendered power dynamics of society and to respond critically to it in a profeminist way. Also, of course, as the pioneer of a chemical-hormonal treatment for sex offenders, it is hard to avoid the conjecture that Money's investment in maintaining the hormonal origin of paraphilia, and the necessity of a concomitant chemical "cure," is driven by a certain amount of self-interest.

The article concludes by presenting a dissenting scientific opinion from Money's that is rooted precisely in an awareness of how rape is not the fantasy of the hormonally/paraphilically imbalanced few, but a characteristic of patriarchal culture more broadly. Miriam Saphira, a Justice Department psychologist and researcher who worked with violent prison inmates, is quoted as finding Money's view inconsistent with her observations: "Rapists channel their anger through sexual arousal. They become hyped up about something and then make moralistic judgments about the woman next door." Saphira makes the valid argument that criminal behavior is an extension or exaggeration of common cultural attitudes, taken beyond the realms of lawfulness, rather than wholly separate from them. Bronwen Reid concludes by reporting: "but in a society that still condones rape in marriage, Ms. Saphira said a rape culture will exist because women are still men's property." This is the kind of perspective that, for all his gestures toward the importance of taking an integrated view of social, psychological, and biological factors, Money seems unwilling ever to consider.

In the example of this newspaper article, we see how, in veering constantly between two etiological explanations of paraphilia that map loosely onto nature and nurture, and using each to bolster his authority when it is convenient to do so, John Money fails to arrive at a consistent, holistic explanation of the conditions of formation of paraphilia. "Bodymind" is integrated only at the linguistic level, as a compound noun, not as a conceptually workable scientific idea. What is especially notable is that Money's, however unsuccessful, obsession with overcoming the dualisms of body/mind and nature/nurture is not coupled with a desire to overcome the equally dubious dichotomy of scientific fact/critical humanities analysis. When he writes in his 2002 lecture that social constructionism belongs to "social science and the humanities where it will be used to explain shifts in sexological ideology," he is describing pretty much exactly what we are doing in this book. However, in so many ways Money's attempts to overcome the historical binarisms characterizing explanations of perversion/paraphilia could have been strengthened by an adoption of such a perspective—one that examines the workings of normalizing social power—as a feature of his method. Where he does turn his energies to deconstructing social ideologies, he only ever targets religion, or sometimes radical feminism, two discourses he tends to run together as the dual enemies of enlightened, objective sexual science (creating yet another unhelpful dichotomy). A serious engagement with the Foucauldian idea that the authority disciplines of psychiatry and sexology are players with power and knowledge (much like religion), and that they construct subjects such as "the paraphiliac" as effects of their games, allows us to apprehend most clearly what is really at stake when Money and his colleagues theorize about the causes and meanings of "abnormality."

NOTES

1. John Money, "History, Causality and Sexology," 2002, typescript of original lecture, 3. (Typescript consulted in the Money Collection of the Kinsey Institute, Bloomington, IN, in March 2011.)

2. Money, "History, Causality and Sexology," typescript, 3.

3. Correspondence between John Money and Eugene Brody, editor in chief, *Journal of*

*Nervous and Mental Disorders*. EB to JM, January 7, 1969. My italics. (All cited correspondence was consulted in the Money Collection of the Kinsey Institute, Bloomington, IN, in March 2011.)

4. John Money and Margaret Lamacz, *Vandalized Lovemaps: Paraphilic Outcomes in Seven Cases of Pediatric Sexology* (Buffalo, NY: Prometheus, 1989).

5. Building on the work begun in John Money, *Lovemaps: Clinical Concepts of Sexual/Erotic Health and Pathology, Paraphilia, and Gender Transposition in Childhood, Adolescence, and Maturity* [1986] (Buffalo, NY: Prometheus, 1988).

6. For a nuanced and thoroughly researched account of the ways in which degeneration theory develops and operates in different European contexts, and is transmitted across them as a major feature of the nineteenth-century worldview, see Daniel Pick, *Faces of Degeneration: A European Disorder c. 1848–1915* (Cambridge: Cambridge University Press, 1989).

7. Money and Lamacz, *Vandalized Lovemaps*, 20–21.

8. Money and Lamacz, *Vandalized Lovemaps*, 21–22. Money seems to mean by these terms the following: "hereditary determinism"—the capacity for abnormal traits to be passed from one generation to another, and for physiological inherited abnormality to be paralleled by behavioral, often criminal, aberrations; "phylogenetic determinism"—the species's programmatic predetermination toward both (apparently "natural") male/female behaviors and the instinct to procreate; "neuropathological determinism"—the likelihood that hormonal or neurological abnormalities will causally coexist with abnormal sexual behavior in an individual; "associative determinism"—the idea that paraphilia is determined by the associations formed between given stimuli/objects and pleasurable, arousing sensations; "intrapsychic determinism"—most closely associated with psychoanalysis, the idea that sexuality is formed by the individual's psychic response to various developmental milestones, most importantly the Oedipus complex; and, finally, "biographical determinism"—observation of the individual's sexual development, whether normal or abnormal.

9. Money and Lamacz, *Vandalized Lovemaps*, 22.

10. Money, "History, Causality and Sexology," typescript, 6–7.

11. Money and Lamacz, *Vandalized Lovemaps*, 17. See also Dany Nobus, "Locating Perversion, Dislocating Psychoanalysis," in *Perversion: Psychoanalytic Perspectives/Perspectives on Psychoanalysis*, ed. Dany Nobus and Lisa Downing (London: Karnac, 2006), 3–18, 6.

12. Money and Lamacz, *Vandalized Lovemaps*, 17.

13. Arnold I. Davidson, "How to Do the History of Psychoanalysis: A Reading of Freud's *Three Essays on the Theory of Sexuality*" [1987], in *The Emergence of Sexuality: Historical Epistemology and the Formation of Concepts* (Cambridge, MA: Harvard University Press, 2001), 66–92.

14. See Nobus, "Locating Perversion," 6.

15. I will explore below, and in chapter 6, the ways in which Money does not wholly reject this logic, even as he nominally adopts a tolerant and "sex-positive" view toward bisexuality and homosexuality. On this logic of sexuality, see also Harry Oosterhuis, *Stepchildren of Nature: Krafft-Ebing, Psychiatry and the Making of Sexual Identity* (Chicago: University of Chicago Press, 2000).

16. Georges Canguilhem, *Le Normal et le pathologique* (Paris: Presses Universitaires de

France, 1966). See Peter Cryle, "The Average and the Normal in Nineteenth-Century French Medical Discourse," *Psychology and Sexuality* 1, no. 3 (2010): 214–25.

17. Vernon. A. Rosario, "On Sexual Perversion and Transsensualism," in *Perversion: Psychoanalytic Perspectives/Perspectives on Psychoanalysis*, ed. Dany Nobus and Lisa Downing (London: Karnac, 2006), 323–42, 328.

18. Money and Lamacz, *Vandalized Lovemaps*, 20.

19. Money's second recommendation is that "counseling . . . needs to be provided for the siblings of a baby with a birth defect; or the offspring or spouse of a paraphiliac sex offender about to be put in prison. . . . Overall, these examples fall into the general category of family therapy, couple therapy, and preventative psychiatry (which includes sex education). At the present time, it is not possible to get third party payments for preventative services. This is in large part due to the fact that *DSM-III* does not recognize these services, and does not have a diagnostic designation for the patients who receive them." Correspondence between John Money and Robert L. Spitzer, MD, chair, Working Group to Revise *DSM-III* (to *DSM-III-R*). JM to RS, July 18, 1983.

20. Money, *Lovemaps*, xvii.

21. Michel Foucault, *The Will to Knowledge: The History of Sexuality*, vol. 1 [1976], trans. Robert Hurley (Harmondsworth: Penguin, 1990), 18.

22. Richard von Krafft-Ebing, *Psychopathia Sexualis with Special Reference to Contrary Sexual Instinct: A Medico-Legal Study*, translation of the 7th German edition by C. G. Chaddock (Philadelphia and London: F. A. Davis, 1920), 1.

23. This introduction is called in different major English translations of the German text "A Fragment of a Psychology of the Sexual Life" (translation of the seventh edition by Chaddock) or "Fragments of a System of Psychology of Sexual Life" (translation of the tenth edition by Rebman). Krafft-Ebing revised the manual several times over a number of years, adding to the later editions increasing numbers of case studies and new subcategories of sexual variation.

24. Krafft-Ebing, *Psychopathia Sexualis*, 1

25. Oosterhuis, *Stepchildren of Nature*, 13

26. Krafft-Ebing, *Psychopathia Sexualis*, 203.

27. Money, *Lovemaps*, 169.

28. Money, *Lovemaps*, 169.

29. Although *Lovemaps* does not return the highest number of hits in a web search of John Money's book titles, this work and its eponymous concept have infiltrated pervasively the popular imagination in the Anglo-American world, for example, via the writing and broadcasting of UK psychologist Oliver James and disgraced psychiatrist Raj Persaud and, in the United States, the advice columnist Dr. Joyce Brothers.

30. John Money, *Sin, Science, and the Sex Police: Essays on Sexology and Sexosophy* (Amherst, NY: Prometheus, 1998), 101.

31. Money, *Lovemaps*, xvi.

32. Money, *Lovemaps*, 267.

33. Money and Lamacz, *Vandalized Lovemaps*, 30–31.

34. Relatedly, Money writes that the antiandrogen drug he prescribed for sex offenders "promises also to be helpful in the regulation of violent temper outbursts in some temporal

lobe epileptics." See John Money, "Determinants of Human Gender Identity/Role," in *Handbook of Sexology*, ed. John Money and Herman Musaph (Amsterdam: Elsevier/North Holland Biomedical, 1977), 57–79, 71.

35. Sigmund Freud, "The Sexual Aberrations," *Three Essays on the Theory of Sexuality* [1905], *Penguin Freud Library*, vol. 7, *On Sexuality*, trans. James Strachey (Harmondsworth: Penguin, 1991), 45–87, 74.

36. Freud, "Sexual Aberrations," 74–75.

37. Foucault, *Will to Knowledge*, 119

38. Money, "History, Causality and Sexology," typescript, 6.

39. Money and Lamacz, *Vandalized Lovemaps*, 39.

40. Money, "History, Causality and Sexology," typescript, 6.

41. John Money, *The Psychologic Study of Man* (Springfield, IL: Charles C. Thomas, 1957), 4.

42. Money, *Psychologic Study of Man*, 6.

43. Freud makes this claim in "The Ego and the Id" [1923], *Penguin Freud Library*, vol. 11, *On Metapsychology*, trans. James Strachey (Harmondsworth: Penguin, 1991), 339–407, 366. This notion, that the ego is a mental projection of the surface of one's body, led to Didier Anzieu's theorization of "Le Moi-Peau" (the skin-ego), as laid out in the book of that name of 1987.

44. Robert J. Stoller writes of Freud: "A fine observer, perhaps the greatest naturalist of human behavior ever, he was also at least as much enthralled by biological speculation. He wished to bridge the gap between the findings of biology, both experimental and natural, and that mysterious product of neurophysiology, the mind." In *Perversion: The Erotic Form of Hatred* [1975] (London and New York: Karnac Books, 1986), 13–14. The most influential book on Freud's indebtedness to biology is probably Frank J. Sulloway, *Freud, Biologist of the Mind: Beyond the Psychoanalytic Legend* (New York: Basic Books, 1979). Sulloway's argument with regard to Freud and biological theory is analogous to my current argument with regard to Money and the logic of nineteenth-century perversion theory: each scientist can be demonstrated to retain debt to the discourse/discipline he disavows.

45. John Money, "Masochism: On the Childhood Origin of Paraphilia, Opponent-Process Theory, and Antiandrogen Treatment," *Journal of Sex Research* 23, no. 2 (May 1987): 273–75.

46. Money, "Masochism," 273–74.

47. Sigmund Freud, "A Child Is Being Beaten" [1919], *Penguin Freud Library*, vol. 10, *On Psychopathology*, trans. James Strachey (Harmondsworth: Penguin, 1991), 159–93.

48. Sigmund Freud, "Fetishism" [1927], *Penguin Freud Library*, vol. 7, *On Sexuality*, trans. James Strachey (Harmondsworth: Penguin, 1991), 345–57.

49. Money, "Masochism," 274.

50. Money, "Masochism," 274.

51. Money and Stoller's names are linked by Richard Green in a paper entitled "The Three Kings: Harry Benjamin, John Money, Robert Stoller." Green writes: "I am the only academic to have worked closely with three sexological kings: Harry Benjamin, John Money, Robert Stoller. These three mentors moulded my career." *Archives of Sexual Behavior* 38 (2009): 610–13, 610.

52. Stoller, *Perversion*, xi.

53. Stoller, *Perversion*, 4. My italics.

54. Stoller, *Perversion*, 6.

55. Stoller, *Perversion*, 109. Stoller's conservatism is noted by Dorothy Allison in a review of his later books, *Observing the Erotic Imagination* and *Presentations of Gender*: "More than any other theorist on this subject, he articulates a reactionary response to radical developments in sexology and gender theory, particularly those that challenge traditional concepts of male and female behavior." "Robert Stoller Perverts It All for You," *Voice*, August 5, 1986, 44–45, 44.

56. "It is important to note by its absence any indication that S/M practitioners have any common psychopathology or symptoms." Charles Moser, "The Psychology of Sadomasochism (SM)," in *SM Classics*, ed. Susan Wright (New York: Masquerade Books, 1999), 47–61, 49.

57. Money, *Sin, Science and the Sex Police*, 60.

58. John Money, Russell Jobaris, and Greg Furth, "Apotemnophilia: Two Cases of Self-Demand Amputation as a Paraphilia," *Journal of Sex Research* 13, no. 2 (1977): 115–25.

59. "One was unmarried, bisexual, and a libertine (John)." Green, "Three Kings," 613.

60. The history of the relationship between psychoanalysis and homosexuality is a fraught one. See Tim Dean and Christopher Lane, eds., *Homosexuality and Psychoanalysis* (Chicago: Chicago University Press, 2001), and Arnold I. Davidson, "How to Do the History of Psychoanalysis." Crucially, Freud himself had argued for allowing gay people to train as analysts, since there was no evidence that homosexuals are more neurotic than heterosexuals. This was a position many of his followers, especially those working in France in the Lacanian tradition, would go on to ignore (to the present day). With regard to Stoller, I have always found it ironic that one of the most hateful accounts of "perversion" in print projects the characteristic of hatred onto its subject matter in its very title.

61. Stoller, *Perversion*, xvii.

62. Stoller, *Perversion*, xvii–xviii.

63. Stoller, *Perversion*, 207.

64. Correspondence between John Money and Maurice Yaffé, principal clinical psychologist, Guys Hospital, London. JM to MY, April 20, 1988.

65. Money, *Vandalized Lovemaps*, 13.

66. Stoller, *Perversion*, 219.

67. Stoller, *Perversion*, ix.

68. Stoller, *Perversion*, xii.

69. Stoller, like Money, was working on the intersection of sex and gender. Money's original formulation of gender was "gender role"; Stoller distinguished "role" from "identity," and developed a concept of "gender identity" in 1968. Robert J. Stoller, *Sex and Gender: On the Development of Masculinity and Femininity* (New York: Science House, 1968). Money then borrowed "gender identity" back from Stoller, while also frequently critiquing the separation Stoller had made between role and identity. Stoller's work on "hermaphroditism" diverged from Money's in the respect that, while Money and the Hampsons at Johns Hopkins popularized the notions that gender was the result of acculturation, Stoller "perceived a 'biological force' behind gender identity." (See Richard Green, "Robert Jesse Stoller, 1924–1991," *Archives of Sexual Behavior* 21, no. 4 [1992]: 337–64, 338.) Yet, despite the reliance on biological standards of normality, Stoller includes the following in his definitions: "by aberration here I mean

an erotic technique or constellation of techniques that one uses as his complete sexual act and that differs from his culture's avowed definition of normality." Stoller, *Perversion*, 3, my italics.

70. Money, "History, Causality and Sexology," typescript, 8.

71. Bronwen Reid, "Men Rape for Sex not Power—Doctor," *New Zealand Times*, December 18, 1983, n.p. (NB: I found this newspaper article in a file of clippings collected by John Money at the Kinsey Institute with the title "Paraphilia." The page numbers had been removed from the newspaper clipping, and I have been unable to source another copy to verify them.)

CHAPTER 3

# Gender, Genitals, and the Meaning of Being Human

*Iain Morland*

John Money's role in the development of infant surgical protocols for the treatment of intersex has been roundly critiqued, but rarely contextualized. His publications on the topic between the mid-1950s and 1970s were singularly influential in shaping medical theory and practice. The speed at which Money's ideas cemented into clinical conventions, as well as their appeal in disciplines such as social psychology and women's studies, appears remarkable. That reception might also seem irreconcilable with later criticisms that infant surgery for intersex is a patently inhumane practice—performed prior to an individual's ability to consent, with few or no functional benefits, and foreclosing self-determination via the imposition of whichever gender can be most easily surgically assigned. The challenge for critical theory, then, is to understand why Money's treatment recommendations ever appeared to be self-evidently right.

My argument in this chapter will be that intersex genital surgery is a humanist enterprise, and that its humanism has been fundamental to the establishment of surgery as a medical protocol. Specifically, I will suggest that the notion of human genitals and gender as surgically and socially plastic depends on the conceptualization in twentieth-century science of plasticity as a quintessential human attribute. In other words, I will argue that Money theorized genitals and gender as malleable at a historical moment when plasticity and humanity were held by Western science to be equivalent. This had the mutually reinforcing effects of facilitating the uptake of Money's ideas about how to treat intersex, while instituting gender as a core human quality, flexible by definition.

I will organize my analysis around Money's test case for intersex treatment—the story of David Reimer. Born male in 1965, Reimer was apparently successfully raised female from 1967, under Money's guidance. In traditional readings of the case, surgery altered Reimer's genitals while parenting altered his gender: the case was thereby reliant on, and demonstrative of, human plasticity. In turn, the case was used to validate intersex treatment protocols, because it seemed to affirm Money's claims that "all the human race" are "psychosexually neutral or undifferentiated at birth," and also that "psychosexual differentiation," when it does occur, derives "ultimately from the genital appearance of the body."[1] Such principles underpinned Money's recommendation that intersex should be treated with the surgical construction of an unambiguous genital anatomy, and the social construction of a psychologically unambiguous gender. But one could believe the principles of neutrality at birth, and genital importance, without accepting Money's treatment recommendations: one might say simply that in the face of intersex, there is nothing to be done. Treatment only makes sense if humans are figured, additionally, as plastic enough to receive it.

*

In Winnipeg, Canada, on August 22, 1965, identical twins named Bruce and Brian were born to Janet and Ron Reimer. Each child had XY chromosomes, testes, and a penis with a full-length urethra: in other words, neither boy had an intersex anatomy. At the age of eight months, on April 27, 1966, Bruce and Brian were taken for circumcision to alleviate phimosis—a narrowing of the foreskin that causes soreness. An accident with the electrical cauterization machine intended to remove Bruce's foreskin caused his penis to be burned.[2] In the following days, it deteriorated to nothing. Crucially, this made Bruce anatomically equivalent to intersex infants born with no penis, a condition called aphallia. More broadly, the accident's outcome presented a predicament comparable to many other cases of atypical genitalia in infancy. What Bruce's situation shared with such cases was the abrupt disruption of social expectations about sex anatomy, together with the discomfiting illumination of usually implicit assumptions about an infant's future sexual function. I will argue that in Canada in the late 1960s, the suppositions of Bruce's caregivers about the relations between sex anatomy and sexual function expressed a tension between mainstream conservatism and new norms of human sexuality. Those norms were publicized by recent sex research, and by contemporary discourse about transsexuality. As I will explain, John Money was involved in both.

Phalloplasty was crude and experimental in the 1960s, so a consultant urologist described penile reconstruction as "out of the question" in Bruce's case.[3] The fact that it was very difficult to remake a penis need not necessarily have been disastrous; it might have been the occasion for pragmatic reflection on sexual function not centered on penile anatomy. But according to initial medical commentators, the irreparability of the accident destined Bruce to a life away from society. This was a problem of deficient copulation, not simply between their aphallic patient and any specific partner, but with the entire realm of social and sexual relations. Following the accident, a local psychiatrist's prognosis for Bruce was that

> he will be unable to live a normal sexual life from the time of adolescence: . . . he will be unable to consummate marriage or have normal heterosexual relations, in that he will have to recognize that he is incomplete, physically defective, and that he must live apart.[4]

The proscriptive, even melodramatic, tone of this report marks a failure of sexual imagination. However, it is not simply one person's prejudice; rather, it is indicative of a national and historical context. The idea of "a normal sexual life," invoked with apparent ease by the psychiatric report, had been challenged profoundly by Alfred Kinsey's best-selling study of male sexual behavior published eighteen years previously. Kinsey famously assumed that, unless interviewees specified otherwise, "everyone has engaged in every type of activity," including numerous activities unrelated to penis-vagina penetration.[5] Although Kinsey's participants were American, his book was read and widely discussed by Canadian clinicians. For instance, in the year of the book's publication, the editor of the *Canadian Medical Association Journal* praised Kinsey for a "scientific approach" that was "free of bias"; and in 1962 and 1964, other authors in the same journal cited Kinsey's work as evidence that mainstream public morals did not reflect the diversity of private sexual behavior, even within what Bruce's psychiatrist would later call "normal heterosexual relations."[6] The psychiatrist's report is interesting, then, not because it evidences a context in which sexual activities other than penis-vagina penetration were flatly unthinkable, but because it shows that even though they were thinkable, they were not considered among the possible futures for a male infant with no penis.

In February 1966, two months prior to Bruce's circumcision accident, Toronto-based psychiatrist Stephen Neiger addressed a sex education symposium for his colleagues from around Ontario on the topic of "recent trends in sex research." He was able to assume that his audience was familiar with Kinsey's "monumental" work.[7] Neiger argued that recent research had improved understanding of "human sexual behaviour," but nonetheless had not affected "university curricula of professions involved in judging norms and offering help in the area."[8] The comments made by Bruce's psychiatrist reflect this: he surely would have known about Kinsey's claims, yet evidently was unable to translate those new sexual norms into clinical practice. Neiger, meanwhile, was dispirited by not only the limited dissemination of sex research among professionals. He observed too that lay understandings of sexual function were outdated—especially regarding the erroneous "question of clitoral versus vaginal orgasm."[9] Arguing that clitoral stimulation was paramount to female sexual pleasure, Neiger cited Kinsey's 1953 book about women, as well as subsequent work by William Masters and Virginia Johnson on sexual physiology.[10] What Kinsey's interviewees had reported, and Masters and Johnson's experiments ostensibly corroborated, was that penis-vagina penetration was inessential to women's pleasure. This implied that aphallia would be no sexual crisis. In fact, Neiger suggested, the quest for "vaginal orgasm" led to malign "artificial symptoms" among his patients.[11] He therefore quoted enthusiastically from a 1965 anthology, *Sex Research: New Developments*, which reprinted not one but two texts by Masters and Johnson regarding clitoral physiology and female sexual response.[12] The anthology's editor was named by Neiger as "another outstanding worker in the field"—John Money.[13] But as I will show, when handling the Reimer case, Money would display the same lack of sexual imagination as Bruce's Winnipeg psychiatrist.

Ten months after the circumcision accident, the Reimers were watching a current affairs television program on which Money was discussing the adult sex change surgeries being pioneered at Johns Hopkins Hospital.[14] The pivotal role of television in the case has rarely received attention. Even while contemporary surgical technology foreclosed some possibilities for Bruce, the technology of television opened others; by the time the Reimers were watching Money in the late 1960s, television had become more popular than newspapers as a source of national and international news for Canadians.[15] Indeed, the program on which Money appeared, *This Hour Has Seven Days*, ran for only two years, but in that time steered the national news agenda on around twenty occasions.[16] Although Reimer's biographer would subsequently suggest that Money outwitted an outraged program host, *This Hour Has Seven Days* was famous for its sensationalism, calculated to win audiences away from competing drama programs.[17] As if in response to Bruce's medical reports, the surgical sex change techniques developed by Money's team were presented in the program as indivisible from their humanistic ethical utility in assisting individuals who might otherwise have to "live apart." Eschewing psychotherapeutic treatments for transsexuality, Money defended surgery as "thoroughly justified in an attempt to constantly increase our ability to help human beings."[18] He thereby invoked, and borrowed authority from, a long-standing popular and professional discourse about medicine as the most humanist of all the sciences.[19]

More directly, Money's comment epitomized late 1960s medical debates about transsexuality, the clinical management of which was regarded by its advocates as mutually expressive of doctors' humanism and their patients' humanity. Such "ameliorative treatment," Money stated in a late 1960s book, "made life a little easier for the patients and, indirectly, for society also"—a claim intended simultaneously to align the treatment process with a mainstream medical agenda of social improvement and to assure critics that post-operative transsexual individuals were more, rather than less, normal than their peers who had not received treatment.[20] This rhetoric, like Money's comments on *This Hour Has Seven Days*, drew a parallel between transsexual individuals and their doctors: just as receiving treatment was claimed to make the former more normal, the treatment provided by the latter testified to their status as normal clinicians, rather than maverick cosmetic surgeons, as some critics had alleged.[21] The discourse of humanism was strategically useful in that context, because it was precisely humanity that transsexual people and doctors could claim to share.

Writing two years after the television program in the foreword to Money's anthology *Transsexualism and Sex Reassignment*, transsexual man and Johns Hopkins benefactor Reed Erickson would describe transsexuality as "a human problem since the most ancient times of which we have any knowledge."[22] Solving this problem through access to surgery, along with counseling and peer support, was one of the ways in which Erickson sought to unlock what he called "human potential," blending the rhetoric of 1960s human rights movements with that of countercultural self-development (which Erickson also supported by funding research into telepathy and hypnosis).[23] Reflexively, Erickson defined the medical management of transsexuality as a uniquely human endeavor:

> The human individual has a basic need to contribute something to the society of which he is a part. Few pleasures can be greater than that of being

> able to alleviate suffering by helping people to fit comfortably into a society that previously had rejected them. . . . All of these rewards have been ours in working with transsexual patients.[24]

In this view, the practice of sex change surgery integrated clinicians and transsexuals alike into society. Accordingly, a happy postsurgical transsexual woman appeared on the 1967 television program to testify, in unknowing parallel to Bruce Reimer's psychiatric report, that before surgery she "was never complete."[25]

When Money went on to state in the program that similar surgeries were being performed at his university hospital on intersex babies with "unfinished genitals," Bruce's parents began to realize that there was an alternative to a "life apart" for their child.[26] Although the clinicians whom the parents consulted had mentioned Money's work with intersex children, the Reimers were unconvinced of its applicability to Bruce until they saw the psychologist on television.[27] Money put to the audience his view that a child with atypical genitalia could be raised successfully in either sex, provided that appropriate surgical techniques were employed. In Money's discourse, transsexual and intersex individuals were both "unfinished" prior to body modification. But if there were technical similarities between surgeries for transsexual adults and those performed on intersex infants, there was also a vital difference in their respective therapeutic aims. Whereas the former were changes of sex to match the category in which patients felt they already belonged, the latter were tools for the assignment of gender in the first place. In the case of transsexuality, then, an adult patient's felt gender was assumed to precede surgery; in the case of intersex, surgery was claimed to create an infant patient's feeling of belonging to a gender. So genital plasticity was supposed in both cases, but only in the case of intersex was gender plasticity equally important. Money's message in the February 1967 television program, which prompted the Reimers to write and request an appointment, was that completed genitals of *either* sex could confer on an "unfinished" infant a flourishing life. However, such a change could make sense only if humans were defined as sufficiently plastic to accept it.

*

By 1967, Money's theory of gender and genital plasticity was already established as a singularly influential and enduring contribution to the clinical management of infants with injured or intersex genitalia. To understand why it could ever have seemed humane to surgically reassign the gender of infants such as Bruce Reimer, the inception of Money's theory in his mid-century doctoral studies must be situated alongside contemporary scientific efforts to define humanity. At that time, it was race, not gender, which was most salient in scientific debates about the meaning of being human. While Money was researching intersex for his PhD during 1950–52, the United Nations Educational, Scientific and Cultural Organization (UNESCO) issued two important statements on human nature in relation to what the organization called "the concept of race."[28] The statements aimed to counter with science any post–World War II vestiges of irrational "racial doctrine" that stood, according to UNESCO, "in glaring conflict with the whole humanist tradition of our civilization."[29] The first statement was written in 1949 and published the following year; authored by leading scholars from anthropology, psychology, and sociology, it went on to be "printed in a considerable number of newspapers in over 18 countries," in addition to receiving extensive radio and television coverage.[30] The second statement was composed in 1951 by a group of anthropologists and geneticists, and similarly published the year after. The fact that UNESCO was quick to revise its initial statement—and that even the amended statement would be appended with invited, worldwide responses by sixty-nine scholars debating almost every assertion within it—shows that the meaning of humanity was a core issue in the academic climate within which Money began to theorize the plasticity of gender and genitals.[31]

Revealingly, the lesson of the UNESCO statements was that the most evolutionarily important characteristic of *Homo sapiens* is "plasticity."[32] Presented as a cutting-edge insight applicable across disciplines, the emphasis on plasticity was employed to trump the notion of "pure race."[33] The latter was implied by both statements to be not just morally objectionable, but deficient as an epistemological tool. As science critic Donna Haraway later observed, for UNESCO and its consultant scholars, "To cast group differences typologically was to do bad science." Instead, "The concept of the population was in the foreground," because it enabled human diversity to be understood as the natural, variable, and contingent outcome of evolutionary processes, rather than the dilution of an original racial hierarchy.[34] Endorsing "wholeheartedly" UNESCO's second statement, influential American biologist Ernst Mayr singled out for special praise its message that "all so-called races are variable populations."[35] Ironically, this postwar, postrace science was nevertheless inextricable from World War II military research, which provided the mathematical models behind the ecological analysis of populations—and which, as I will explore in chapter 4, enabled ways of thinking predictively about complex systems that Money too claimed to use.[36]

Population variability was personalized by UNESCO as individual educability. Just as the population concept allowed the organization to say that "genetic differences are of little significance in determining the social and cultural differences between different groups of men," so too were differences between individuals described as predominantly autonomous from their anatomy.[37] This was in line with contemporary biological science, which, inspired by wartime systems analysis, conceptualized individual learning and group evolution as the same mechanism working on different scales—and which held, further, that a profound capacity for education differentiated humans from animals.[38] "The normal individual, irrespective of race, is essentially educable," stressed UNESCO's 1952 statement; "It follows that his intellectual and moral life is largely conditioned by his training and by his physical and social environment."[39] Shortly afterward, in 1955 and 1956, Money and his clinical mentors would publish six key articles in which gender would be formulated in the same manner: based on their work at Johns Hopkins with seventy-six intersex patients, Money and colleagues claimed that "sexual behavior and orientation as male or female does not have an innate, instinctive basis."[40] Put differently, humans could be educated into either gender. Specifically, "a person's gender role and awareness is founded in what, from infancy onward, he learns, assimilates and interprets about his sexual status from parents, siblings, playmates and others, and from the way he reads the signs of his own body," they stated.[41]

Nonetheless, there might seem to be a divergence between UNESCO's view and a clarification by Money and his colleagues that theirs was not "a theory of environmental and social determinism." But there was unmistakable common ground, because the basis for the Money team's rejection of determinism was that "transactions are frequently highly unpredictable, individualistic and eccentric."[42] This idea that complex systems, resistant to predetermination, could generate outcomes of simple and stable appearance, was precisely the science that UNESCO embraced in order to debunk "race." In the same way that "myriad experiences encountered and transacted" were held by Money and his colleagues to create a gender role as either "boy or man, girl or woman," so did UNESCO assert that "the Mongoloid Division, the Negroid Division, [and] the Caucasoid Division" were really "dynamic, not static" processes, momentarily "embalmed" into classifications.[43]

I think that the congruence between Money's claim about gender and contemporary scientific debates about race gave his work a self-evidence that was crucial to its broader uptake. The foundational publications appeared in the relatively minor *Bulletin of the Johns Hopkins Hospital*, but the theory expounded by Money and his colleagues reached with extraordinary speed a receptive wider audience. This has been overlooked in some reviews of their work.[44] Having dispatched in under two months their first five papers to the in-house *Bulletin*, in 1956 the authors industriously repackaged their findings, with a genetic psychology slant, for an international pediatrics conference in Copenhagen (published a year later in the *American Medical Association Archives of Neurology and Psychiatry*) and as a "teaching clinic" paper about intersex treatment in the *Journal of Clinical Endocrinology and Metabolism*.[45] This shows that from his earliest writings on intersex, Money sought to have a multidisciplinary impact on both theory and practice. At a June 1956 meeting of the American Association of Genitourinary Surgeons, Joseph H. Kiefer, a surgeon from the University of Illinois, praised the Money team's "very excellent recent studies" of "psychic factors" in the gender development of individuals with intersex anatomies. Subsequently published in the American Urological Association's official journal, Kiefer's presentation was another key early vector for the dissemination of Money's theory to clinicians nationwide.[46]

The most authoritative commendation came from the leading pediatric endocrinologist Lawson Wilkins. Between the first edition of Wilkins's respected textbook *The Diagnosis and Treatment of Endocrine Disorders in Childhood and Adolescence* in 1950 and the second edition in 1957, Money and his colleagues had joined Wilkins's clinic at Johns Hopkins, where they studied the patients who allegedly provided the basis for their theory about gender. In his textbook's second edition, which would be reprinted twice within five years, Wilkins inserted the claim that "the gender in which a child is reared is the predominant factor in determining the future psychosexual orientation," citing the first five *Bulletin* papers as supporting evidence.[47] Wilkins had already, in a 1955 *Pediatrics* article, made this point with reference to a *Bulletin* paper that was currently in press.[48] But in the textbook, he went further, turning over to Money and his colleagues a page of the chapter on sexual precocity, where they reproduced (with minor edits) treatment guidelines that they had recently published in *Psychosomatic Medicine*, together with their signature claim that "psychosexual maturation is determined by various life experiences encountered and transacted and is not predetermined as some sort of instinctive or automatic product of bodily sexual maturation."[49] It was in an effort to describe exactly this difference between the body and "psychosexual maturation" that Money had coined the term "gender," as something plastic by definition. By spring 1961, at a meeting of the California Medical Association, the theory of gender plasticity was being discussed by psychologists as an established fact; and when a medical team from London's Great Ormond Street Hospital cited several *Bulletin* papers in a 1963 issue of the *Lancet*, they described Money's account of gender as "currently accepted," along with its implications for treatment.[50]

When the Reimers took Bruce to Johns Hopkins in spring 1967, Money used the theory that gender "does not have an innate, instinctive basis" to explain that their son could be raised successfully as a girl, provided that reassignment surgery was carried out promptly.[51] His conviction was that the gender of any child remained malleable until the closure of what he would later call a "gender identity gate" (as noted by Nikki Sullivan in chapter 1) between the ages of around three and four years.[52] The optimal period for reassignment, however, was theorized by Money to be the first eighteen months of life.[53] This was because, as Money had stated in a 1962 publication, "the immature brain has greater plasticity"—or, as he would phrase it ten years later in *Man and Woman, Boy and Girl*, "the human brain at birth is still, embryologically speaking, not finished."[54] The claim that all humans are born cerebrally unfinished provided a foundational justification for studying the effects of surgical reassignment on infants with atypical genitalia. It enabled Money to present such cases as widely relevant, rather than exceptional. However, in his work with intersex infants, Money had been unable to exclude the possibility that their purported acceptance of reassignment might itself be caused by intersex development—in other words, that having an intersex anatomy could entail a predisposition toward plasticity. His work had been criticized on this basis in 1965.[55] The Reimer case therefore interested Money because it was a chance to test the theory of plasticity in an infant without intersex development.

Money would later call the proposed reassignment not merely "the best informed" course of action, but also "the most humane" at the time.[56] The question of Bruce's plasticity, then, was constituted as a matter of professional and personal concern alike. Aged seventeen months, Reimer would need surgery promptly, to ensure a therapeutic outcome for the family, and simultaneously to verify for Money and his peers the gender identity gate theory. Early medical commentators on Money's work, such as Kiefer, had suggested that gender in intersex individuals remained plastic for longer—up to "later childhood and even teen age."[57] So, although in most popular accounts of the case, Bruce's reassignment has been depicted as the occasion for Money to demonstrate gender's unlimited plasticity, it is more accurate to say that Money sought to underscore gender plasticity's temporal limit, while nonetheless assuming the surgical adaptability of genitalia. But since the reassignment could be therapeutic only if the gender theory were true, and the gender theory could be substantiated only by a therapeutic outcome to the reassignment, Money's logic in recommending treatment for Bruce was circular, however "humane" it may have appeared.

*

Shortly after the Reimer family's meeting with Money, Bruce was renamed Brenda, and began to be raised as a girl. On July 3, 1967, Brenda's testicles were removed by Money's colleague, surgeon Howard Jones Jr. At the same time, Jones performed preparatory surgery for the construction of a vagina, reshaping the scrotum to provide the appearance of labia majora.[58] Despite the exceptional circumstances of the circumcision accident, this surgery typified the feminizing procedures for intersex that Money and his colleagues had been advocating for over a decade. Feminization procedures included the reduction of phallic tissue to produce the appearance of a clitoris, in infants born with what clinicians regarded as an ambiguously sized phallus; and castration, in cases where clinicians anticipated that chemicals secreted by the testes would cause confusing physical development at puberty.[59] Wholesale construction of a vagina was usually deferred until an individual was older, as was the intention with Brenda, on the assumption that a fully grown anatomy would be more manageable in surgery—although the same reasoning was not applied when scheduling interventions on the rest of the genitalia. Money's explanation for this disparity was that vaginal absence "is of remarkably little concern to younger girls," becoming "a psychologic issue only in early adult life."[60] His statement is significant because it marks a switch point between surgical technique and psychological concerns. I will argue that just as Money's theory of gender echoed postwar discourse about race, the treatments that followed from the theory—including Brenda's—extended an older discourse about the use of surgery to produce psychological effects. Further, humanism was what these pre- and postwar discourses had in common.

Money neither performed nor invented any surgical procedures.[61] His innovation in the 1950s was to rationalize the use of surgery to create whatever genital appearance would facilitate the necessary "experiences encountered and transacted" in order to produce a gender role of "boy or man, girl or woman." This entailed feminization more frequently than masculinization because, like vaginal construction, masculinizing surgeries in infancy were regarded as technically difficult; but it was also because, unlike being a girl without a vagina, being a boy with a less-than-typically masculine anatomy was considered by Money to be psychologically damaging.[62] Consequently, although both Money and surgeon Jones published recommendations that masculinizing surgery should be delayed until a child was old enough to operate on easily (Jones put the age at four in 1958; Money at five or six in 1961), this advice was trumped by their guideline that surgery ought to forestall self-awareness of atypical genitalia.[63] Since feminization was the only kind of reassignment surgery that could be done sufficiently early in life to have this purported effect—preceding the closure of the gender identity gate—Money's theory of plasticity led to asymmetric clinical practice: feminization became the norm, and masculinization the exception.[64] This was the state of affairs that the feminization of Reimer demonstrated. In that context, the option of raising Bruce as a boy without a penis, deferring surgery until his body had grown and phalloplasty techniques had improved, appeared psychologically hazardous. An ambiguous genital appearance was thought by Money to be detrimental, then, because it created "doubts and perplexities" for the developing self, and for others.[65] In this way, the measure of surgical reassignment was situated by Money in the psychological realm, where, as a psychologist, he was able to claim expertise. Accordingly, he censured clinicians who did not prioritize "examination of the psyche" in intersex treatment.[66]

That the transposition into medical practice of a psychologist's gender theory would mean genital surgery—instead of, for instance, a flexible attitude to the relation between genitalia and gender—requires interrogation. The recommendation of surgery for Brenda contrasted with Money's contemporaneous work with transsexual patients, where he was liberal about the implications of genital diversity. For example, describing the results of penile construction as "a lump of meat," Money nevertheless reported on a postoperative transsexual man who, together with his wife, was "able to have a rather good time sexually," including "the satisfaction of climax" for both parties. However, Money maintained skepticism about the value of painful, time-consuming phalloplasty, encouraging transsexual use of "an artificial penis of the strap-on, dildo variety."[67] These remarks reveal Money's awareness of at least two arguments against feminization in infancy—first, that a man with unusual genitalia could successfully have sexual relations; and second, that use of a prosthesis could circumvent the need for surgery entirely. Money further criticized his transsexual patient for being "obsessed" with wanting to stand to urinate, which was reportedly unfeasible after phalloplasty.[68] But the enablement of a standing position during urination was fundamental to masculinizing surgery for intersex too, and the very reason why masculinization was considered too complicated to do in infancy.[69] There was, then, a failure of imagination by Money in his handling of the Reimer case, which exemplified a difference between his approaches to intersex and transsexuality. Money might have defined a psychologist's role in intersex treatment as fostering tolerance toward genital diversity similar to his own regarding transsexuality; and since five months elapsed between Bruce's renaming as Brenda and the surgery by Jones, Money might have supported the Reimers in continuing to raise Brenda without surgery, even if he thought her gender reassignment was beneficial.

Money's actions followed from assumptions about human nature grounded in his psychology training during the 1940s. By defining gender as plastic, and thereby open for a time to medical construction, Money assumed plasticity to be innate, inevitable, and not a construction. He never acknowledged that plasticity might vary between individuals, or be shaped by experience. A belief in essential human plasticity undergirded Money's claims that children were profoundly educable; that childhood experiences endured; and that any uncertainty generated by genital ambiguity would therefore be perilous because of its lasting impact on gender. While the conclusion about gender was Money's own, the steps by which he reached it were representative of humanist psychology in first half of the twentieth century. Historian Joanne Meyerowitz has shown how, by midcentury, the discourse of transsexuality was inflected by psychologists' exhortations to "self-actualization."[70] Originating with the American Carl Rogers, the popular theory of self-actualization held that humans incline toward goals that provide enhancement and fulfillment, on the basis of sensory and intuitional feedback.[71] Transsexual individuals' requests for medical assistance, including surgery, could be understood in this manner; indeed, some transsexual people wrote to clinicians using the terminology of self-actualization.[72] But Rogers's humanist psychology was inapplicable to intersex treatment. Although Rogers interpreted infant behavior as self-actualization, he thought it rudimentary in comparison to the confidence of a "psychologically mature adult" to make judgments as a "total person," reflecting and remembering; a fully actualizing self was a mature self.[73] Contrastingly, Money advised that treatment for intersex should precede self-awareness and memory formation, taking place when there was barely a self to actualize. Whereas Money theorized natural plasticity—likening treatment to the fashioning of "clay"—Rogers maintained that humans were not "malleable putty which can be shaped into *any* form."[74] Therefore, I argue that Money's ideas must be tracked to a prior humanist psychologist, the Austrian Alfred Adler.

In the early decades of the twentieth century, Adler argued for the psychological primacy of self-defense against "the ravages of the external world."[75] For him, a child's capacity to be affected by its environment was the basis for both sociability and educability, whereby humans learn to influence one another, and the world. Adler contended that the weakness of all children relative to their environment provides a "stimulus of inadequacy" from which a "thousand talents and capabilities" can develop.[76] In his universalizing emphasis on educability, Adler was a seminal forerunner to the midcentury humanism represented by UNESCO. Moreover, Adler provided an account of development on which Money, I think, modeled his own. Just as Money claimed that the "critical growing-up period for establishment of gender role and orientation is reached by the age of about 18 months," so too had Adler suggested that lifelong attitudes were set "in the first months."[77] In the same way that Money theorized the gender identity gate's closure between ages three and four, Adler had proposed that the "character traits and psychic movements" of adulthood are traceable to the "third and fourth year of life."[78] Both men critiqued the explanation of intractable psychological characteristics as hereditary, reinterpreting them as evidence of infantile impressionability.[79] I argue that Money's work on intersex extended this axiom to assert that "psychosexual differentiation" is, inherently and ambivalently, "vulnerable to many interferences."[80] This was important because it connected theory to practice. If atypical genitalia could interfere with psychosexual differentiation, so too could early surgery. Money believed that children's developmental capacity permitted therapeutic "flexibility and malleability," as well as susceptibility to cumulative psychological defects.[81] For Money as for Adler, then, educability and vulnerability were two facets of human plasticity.

It is likely that Money studied Adler as a psychology undergraduate in the early 1940s. By the time Money published on intersex in the mid-1950s, the 1928 English translation of Adler's bestselling *Understanding Human Nature* had been reprinted six times, and Adler's terminology was circulating in the United States.[82] The diagnostic concept for which Adler was most renowned, which entered both everyday and medical discourse, was "inferiority." Simultaneously an existential condition into which all infants were born and a distinct developmental problem, Adler's idea of inferiority linked anatomy to psychology in a way that laid the ground for Money's treatment recommendations. For Adler, humans born with "inferior organ-systems" were unable to meet environmental demands, whether through specific organ defects, mobility difficulties, or general frailty.[83] In a 1949 paper preceding his work on intersex, Money used the phrase "organ inferiorities," demonstrating knowledge of Adler's theory.[84] Crucially, Adler psychologized inferiority as a feeling, reflecting not an individual's "actual inferiority" but rather their "interpretation" of it, however exaggerated.[85] Here, educability reversed into vulnerability: infantile perceptions of the world's unaccommodating hostility developed into childhood "isolation" and persistently diminished sociability.[86] This effect prefigured strikingly the psychiatric prognosis for Bruce Reimer in 1966. But the symmetries between interwar treatments for inferiority and Money's advocacy of gender reassignment surgery are even more striking.

Plastic surgery had advanced in technique and credibility during World War I.[87] Through treating battlefield mutilations, surgeons endeavored to reconstruct not only bodies, but also shell-shocked minds. After the war, plastic surgeons continued to employ psychology in promoting their expertise to wider publics; when the American Society of Plastic and Reconstructive Surgeons was founded in 1931, its objectives included the dissemination of the profession's "psychological importance."[88] Society cofounder Jacques Maliniak was representative in this respect: he used Adler's concept of inferiority to communicate not simply the problems that plastic surgery could cure, but those which it could supposedly prevent. Maliniak, like his colleagues, thought that "inferiority feelings and anti-social tendencies" arose in individuals with characteristics such as a cleft lip or nasal fracture.[89] Despite specifying that these psychological problems affected "hypersensitive individuals," Maliniak stated that the characteristics themselves caused "hypersensitivity," rehearsing Adler's link between organic weakness and mental impressionability.[90] For children, Maliniak regarded the consequences as particularly damaging, including "ridicule" by peers and "persistent awareness of disfigurement."[91] Prompt plastic surgery was therefore recommended by Maliniak to minimize the physical trace of "deformities" and prevent them from ever entering a child's consciousness.[92] He argued correspondingly that surgery ought not to wait until puberty and that the goal of "psychic health" outweighed risks associated with repeated surgery during childhood.[93] In 1935, Maliniak took this message to parents in the popular American health magazine *Hygeia*, claiming ingeniously that parental "distress" about a child's appearance was actually beneficial, because it motivated parents to arrange early surgery.[94] Like Adler, then, Maliniak traced lifelong psychological qualities to childhood; unlike Adler, he transformed this into a justification for plastic surgery.

Maliniak's arguments reveal how, in the decades before Money's work on intersex, plastic surgery was routinely represented as a means to avert inferiority feelings in the developing self. Such discourse continued through the 1950s, when the widely published American plastic surgeon Maxwell Maltz defined blemishes as only those disfigurements of which a person is conscious.[95] A body without blemishes was therefore a mind without blemishes; and an unblemished child was one with no memory of receiving surgery. This important context explains why Money cast the role of a consulting psychologist in intersex cases as not to defend genital diversity, but to oversee its prompt surgical elimination: in doing so, Money took up the established, Adler-inspired imperatives of plastic surgery—to operate early in life, to foreclose self-consciousness of atypical appearance, to prevent teasing, and to reduce parental anxiety. Echoing the title of Maltz's surgical autobiography, *Doctor Pygmalion*, Money called the process of gender reassignment "Pygmalion-like."[96] He innovatively named the measure of treatment's success "gender," but the outcome that Money sought was as derivative as his Pygmalion allegory. Asserting in 1956 that "no person in our society could be other than crippled as an effective and happy human being without a sturdy conviction of belonging either to one sex, or the other," Money expanded the familiar remit of plastic surgery to include enabling a "sturdy conviction" of one's sex, which would have been recognizable to Money's contemporaries as the opposite of inferiority feelings.[97] Couching this project in explicitly humanist terms (producing "effective and happy" human beings) made it seem even more normal to parents and medical professionals. Interlinking the plasticity of the body with the plasticity of the mind, Money followed plastic surgeons in transfiguring human vulnerability into therapeutic malleability. Underpinned by Adlerian psychology, the aim was to protect humans from themselves.

*

After the surgery in 1967, the Reimers brought their daughter to Johns Hopkins annually for monitoring by Money. According to his retrospective account of the case, published in 2002, these follow-up meetings continued until 1974, when Brenda was nine. The family then lived farther away until 1978, when a follow-up was cut short because Brenda was "totally unable to respond to any talk pertaining to sex or sex education," and "fled in panic" from the meeting—although Money did not mention this significant incident until twenty years later.[98] Beginning in 1972 with *Man and Woman*, Money had presented in print and at conferences a series of accounts of the case, based ostensibly on the annual reviews. Appearing in both academic and popular publications, Money's accounts often conflicted over details, but consistently asserted that Brenda's reassignment had been a therapeutic success and that it constituted evidence for his gender theory. As I will show, by the late 1970s, these claims were repeated confidently by scholars in sociology and psychology, apparently regardless of the deficiencies in Money's reports. In the early 1980s, that started to change. But rather than adjudicate between interpretations of the case as demonstrative of gender theory, I will reveal that Money and his leading critic, Milton Diamond, shared a belief in human plasticity.

Even allowing for delays between writing and publication, Money's case reports chart a peculiar chronology, which has not hitherto been scrutinized; commentators have assumed that his reports were internally coherent, albeit misrepresentative. Yet, Money's reports were never coherent. In *Man and Woman*, published in the year Brenda turned seven, Money stated that his patient was "five years and nine months old" at their most recent meeting.[99] This gap might be explained by the duration of the publishing process; however, two publications in the following year reveal a trend to chronological inconsistency. On January 8, 1973, *Time* magazine covered the case, publicizing a talk given a week earlier by Money to the American Association for the Advancement of Science. The *Time* article seemed to be based on an interview with Money, and included a photograph of him at work. The article stated that the circumcision accident happened in October 1963—two years *before* Brenda's birth—and described her as "a nine-year-old child who is, psychologically at least, a girl."[100] At the time, Brenda was only seven and a half. In an academic paper published later that year, Money recalled that the coverage in *Time* prompted letters to him from parents of children born with small penises.[101] But in his paper, a lightly edited transcript of lectures delivered in early 1973 at the University of Nebraska, Money claimed Brenda to be "nearly 9 years old"—younger than stated in the earlier *Time* article and still older than Brenda's actual age.[102] She would turn eight in August 1973. Confusingly, Money illustrated his paper with photographs of Brenda at age six. He suggested that her cross-legged pose, with elbow on knee and hand on cheek, represented spontaneously "feminine" body language—a comment that was at once anachronistic and vaguely patronizing, considering the age of the subject and the wider historical moment of feminist activism and equal rights legislation in the United States.[103]

Although in 1973 Money had looked forward to the induction of Brenda's puberty, by estrogen injections, as the "capstone" of her gender development, two years later in *Sexual Signatures* he rewound to discussing only her first five and a half years of life.[104] Superficially, this can be explained by Money's claim that *Sexual Signatures* was merely a "popularized trade book adaptation" of *Man and Woman.*[105] But despite not updating the information given in *Man and Woman*, the conclusions drawn in *Sexual Signatures* were more forthright than those in the previous book. In *Man and Woman*, Money used the Reimer case to demonstrate "that gender dimorphic patterns of rearing have an extraordinary influence on shaping a child's psychosexual differentiation and the ultimate outcome of a female or male gender identity," and cautioned that "the child is not yet postpubertal and erotically mature, so that the final word remains to be written."[106] In *Sexual Signatures*, contrastingly, he called the case "dramatic proof" and "convincing evidence" in support of the theory that "the gender identity gate is open at birth for a normal child no less than for one born with unfinished sex organs."[107] Even if Money believed it to be true, this assertion was implausible. The outcome of a single case study could not constitute "convincing evidence" by any scientific standard. The fact that *Sexual Signatures* was written for a lay audience does not invalidate this criticism, because Money later chided media commentators for interpreting the Reimer case as "final proof" of gender theory.[108] Criticizing them for drawing conclusions "on the basis of N=1 with no control case," Money overlooked that his own work had referred to Brenda's brother as the "control."[109]

In the same year as *Sexual Signatures*, 1975, a paper by Money appeared in an academic volume about sexuality and psychoanalysis, derived from a talk given to the Society of Medical Psychoanalysts. In it, he made passing reference to the Reimer case, stating simply that Brenda was "now aged seven" and "clearly shows female gender-identity differentiation."[110] But in another 1975 paper, containing a report on the case adapted from the text of *Man and Woman*, Money wrote that Brenda was "now 9 years old," and enthused again about the promise of estrogen therapy, which he stated would allow Brenda to develop a "sexually attractive appearance."[111] I think that Money's persistent lack of chronological clarity is indefensible, given the central importance of the case to his signature theory of plasticity, as well as the acute relevance of the timing of treatment to the purported action of the gender identity gate. In addition to inconsistencies over Brenda's age, Money's accounts were incongruent regarding the date of his first meeting with the Reimers. In *Man and Woman* and the 1975 paper derived from it, as well as a 1998 retrospective, he claimed to have met the Reimers for the first time on the occasion of Brenda's surgery.[112] However, in *Sexual Signatures* and his 2002 retrospective, Money wrote that he met the family before surgery, which accords with Reimer's own biography.[113] Despite publishing profusely throughout his career, Money never provided a comprehensive, logically ordered and dated account of the Reimer case. His final confident statement about the outcome appeared in a 1978 book chapter, where he stated in two short paragraphs that Brenda's gender was "tomboyish" yet "feminine."[114] The chapter featured a full-page photograph of the twins, but its caption did not specify their age.[115] After that, Money published no substantial comments on the case for two decades.

The inconsistencies between Money's accounts did not preclude their widespread uptake by other authors in the late 1970s. Referenced in "virtually all academic writing on sex and gender," according to psychologist Suzanne Kessler, the story of the boy who had been changed into a girl appeared to literalize gender's social construction, and thereby "to have struck a mighty blow to biological determinism."[116] For many feminists, Money's gender theory showed that patriarchal social arrangements were open to change—even though in *Time* magazine, he had concluded conversely that the gender identity gate's closure in childhood rendered futile feminist efforts to change adult attitudes.[117] Contemporary psychology and sociology textbooks, which otherwise expressed feminist caution about sexism in science, presented Money's claims about the Reimer case in often uncritical ways.[118] One described *Man and Woman* as "a well-done presentation of findings"; another reprinted the version of the Reimer case from *Sexual Signatures*, almost in entirety, as a reading for students.[119] *Sexual Signatures* was also "highly recommended" in *Library Journal* "for high school, college, and public libraries," while a reviewer for the recently launched feminist journal *Signs* preferred *Man and Woman* for its "superb account of human sexual development," especially the "startling" reports of gender reassignment.[120] By the close of the 1970s, despite a few keen feminist critiques of Money's assumptions about appropriate gender roles, the influence of his theory extended beyond medicine to the humanities and social sciences.[121] This process, smoothed by the ubiquity of Money and his collaborators on the editorial boards of sexuality studies journals, had been accelerated by the apparent generalizability of the Reimer case's outcome.[122]

However, the outcome was not as Money had claimed. Investigating the case in 1979, a documentary team from the British Broadcasting Corporation (BBC) found Brenda to be unhappy as a girl. Although psychiatrists at the family's local hospital were continuing to treat her as female, and had not told her of the reassignment, in anonymous interviews with the BBC they expressed skepticism about the prospect of Brenda's "adjustment as a woman."[123] Reasons for the psychiatrists' concerns included Brenda's "very masculine gait," her ambition "to be a mechanic," perception that "it's easier to be a boy than it is to be a girl," and insistence, in a stick-figure drawing test, that it was "easier to draw a man."[124] Seeking commentary on these revelations, the BBC interviewed Milton Diamond—the biologist who in 1965 had critiqued the universality of Money's claims about gender, on the grounds that individuals with intersex anatomies might be more plastic than those with typical female or male development. Money too agreed to be interviewed for the documentary, but then withdrew, stating later that the program-makers broke a promise to him that "they would not interview the twin or discuss her case on camera" and that he was "wrongly typecast as the advocate of social rearing to the exclusion of biology."[125] Diamond's own position was that the human nervous system usually differentiates prenatally, in response to hormones, and thereby predisposes individuals toward specific gender roles. He held that the nervous system at birth is therefore biased, "modifiable but not to the extent of complete reversal or negation."[126] For Diamond, this had implications for patient care. Reporting in a 1982 article on what he had learned about Brenda from the documentary, Diamond proposed that the durability of a "male-based" nervous system meant that aphallic children "should be unequivocally raised as males."[127] But this was not an absolute break from Money: the limits of plasticity still determined the meaning of humane treatment.

At the time of his interview, Diamond had no contact with the Reimer family or with Brenda's psychiatrists. In the early 1990s, he succeeded in identifying and contacting Keith Sigmundson, who had managed Brenda's local psychiatric team. Diamond learned that Reimer, having taken the name David, was living unequivocally as a male. Agreeing eventually to meet with Diamond, David was shocked to hear about the widely disseminated 1970s reports of his life and their role in affirming treatments for intersex.[128] With David's consent, Diamond and Sigmundson published in 1997 a detailed follow-up based on pseudonymous interviews during 1994–95 with David and his family. In the interviews, David and his mother recollected that he rebelled from age seven against the annual visits to Johns Hopkins and expressed typically masculine interests throughout childhood. Having refused to live as a girl even before learning of his reassignment in infancy, David underwent elective mastectomy and phalloplasty during his mid-teens, after discovering his past. He began taking testosterone. At age twenty-five, he married and adopted his wife's children.[129]

Following Diamond and Sigmundson's article, David went on to collaborate with the journalist John Colapinto, publishing a biography and publicly revealing his identity in 2000.[130] Meanwhile, Diamond and Sigmundson issued recommendations for the treatment of individuals with injured or intersex genitalia. Expanding Diamond's argument for gender assignment according to the likely bias of an infant's nervous system, they advised that assignment as male or female ought to remain sufficiently "flexible" to accommodate a mature patient's desire to change gender, should it arise.[131] Crucial to such flexibility, in Diamond and Sigmundson's view, was the avoidance of irreversible cosmetic genital surgery on infants and children.[132] This emphasis on flexibility signaled the persistence of humanism in their critique, despite their divergence from Money. Responding to the charge that some recipients of early surgery were nevertheless happy, Diamond and Sigmundson stated that "humans can be immensely strong and adaptable."[133] Remarkably, this was the same point that Joan Hampson, one of Money's earliest collaborators at Johns Hopkins, made in 1955 about individuals with intersex anatomies who had *not* received surgery. Hampson claimed that such people illustrated "the surprising adaptability of the human organism."[134] I argue that the continuity between these otherwise dissimilar accounts, written over forty years apart, shows how twentieth-century humanism has structured both the emergence and the rejection of Money's work.

Money engaged on two occasions with the revelations about the case's outcome.[135] In a 1998 book chapter, he outlined some "possible contributing variables" for David's "second sex reassignment" as male.[136] The majority of these "variables" contradicted Money's earlier claims about the feasibility and outcome of Bruce's reassignment. They included the detail that surgery did not take place "until the advanced age of 22 months"; Brenda's realization that "identical twins are always both boys or both girls"; and Brenda's discovery "in the sexual play of childhood" that her "genital self" was "neither male nor female"—all of which might have been foreseen in 1967, yet which were unproblematic according to Money's 1970s publications.[137] Indeed, in *Sexual Signatures*, Money had stated there to be "nothing" about Brenda's "genital appearance to make her feel self-conscious, even if she gets into the bathtub with her twin brother."[138] Another variable suggested by Money in 1998 was the "existence of a trust fund for only one twin, not both," referring to the compensation paid following the circumcision accident.[139] However, David had no access to that fund until he turned eighteen, by which time he had lived as a male for four years.[140] In a 2002 chapter, Money asserted that no part of Reimer's biography "may be accepted at face value," because Colapinto had mistakenly called the American Psychiatric Association the American Psychiatric Society.[141] That this point would be a cornerstone of Money's counterargument is indicative of the fact that neither of his retrospective chapters underwent external peer review.[142] The later book's manuscript had been rejected by an editor, so was published by Money's former student, the psychiatrist Richard Green.[143] Four years afterward in an obituary of Money, Green recounted that his mentor preferred to withdraw manuscripts rather than to revise them.[144] This might explain why the obfuscations in Money's 1970s case reports were not corrected before publication. It also sheds light on the deficiencies of Money's retrospective chapters. Instead of engaging thoroughly with critiques that had major ramifications for intersex treatment, Money padded his chapters with sections on topics such as "horoscopy and causality," "semen-conservation theory," a four-page quotation from his own doctoral dissertation, and eighteenth-century debates about the composition of the earth's crust.[145]

*

Reimer chose to end his life in 2004. I did not foreshadow this event when introducing his story, because that would have framed his life as "the tragedy of David Reimer," in the words of one commentator.[146] As a narrative structure, tragedy leads inevitably to disaster. In the present context, it would imply infant genital surgery to be a monumental intervention with overpowering consequences. Reimer's life shows that not to be the case. I therefore disagree with a comment by one intersex support group that Reimer's story "teaches us most clearly . . . how much people are harmed by being lied to and treated in inhumane ways."[147] The case can teach us about some of the ethical and scientific shortcomings in Money's work. But in view of my analysis in this chapter, I am unconvinced that we can straightforwardly demarcate inhumane from humane treatment, even though it would be reassuring to do so. I argue that the history of humanism cannot decide the meaning of a humane response to atypical genitalia. In other words, we cannot discover the right way to treat individuals with intersex anatomies by determining what it means to be human. The challenge for medical reformists, then, is to contest the legacy of Money's treatment model not for its dehumanization of patients, but for its assumptions about their humanity.

NOTES

1. John Money, "Cytogenetic and Psychosexual Incongruities with a Note on Space-Form Blindness," *American Journal of Psychiatry* 119 (1963): 820–27, 820; John Money, "Psychosexual Differentiation," in *Sex Research: New Developments*, ed. John Money (New York: Holt, Rinehart and Winston, 1965), 3–23, 11–12.

2. John Colapinto, *As Nature Made Him: The Boy Who Was Raised as a Girl* (London: Quartet, 2000), 9–12.

3. Christopher J. Dewhurst and Ronald R. Gordon, *The Intersexual Disorders* (London: Baillière, Tindal and Cassell, 1969), 85–86; M. Schwartz quoted in Colapinto, *As Nature Made Him*, 15.

4. G. L. Adamson quoted in Colapinto, *As Nature Made Him*, 16.

5. Alfred C. Kinsey, Wardell B. Pomeroy, and Clyde E. Martin, *Sexual Behavior in the Human Male* (Philadelphia: Saunders, 1948), 610–66, 53.

6. H. E. MacDermot, "The Scientific Approach," *Canadian Medical Association Journal* 58 (1948): 505; Anonymous, "The Other Side: Living with Homosexuality," *Canadian Medical Association Journal* 86 (1962): 875–78, 877; H. E. Emson, letter to the editor, *Canadian Medical Association Journal* 90 (1964): 1037.

7. Stephen Neiger, "Recent Trends in Sex Research: New Facts for the Clinician—Horizons for the Psychologist in Research," *Canadian Psychologist* 7a (1966): 102–14, 103.

8. Neiger, "Recent Trends in Sex Research," 113, 103.

9. Neiger, "Recent Trends in Sex Research," 108.

10. Alfred C. Kinsey, Wardell B. Pomeroy, Clyde E. Martin, and Paul H. Gebhard, *Sexual Behavior in the Human Female* (Philadelphia: Saunders, 1953).

11. Neiger, "Recent Trends in Sex Research," 109.

12. William H. Masters and Virginia E. Johnson, "The Sexual Response Cycle of the Human Female: Gross Anatomic Considerations" and "The Sexual Response Cycle of the Human Female: The Clitoris: Anatomic and Clinical Considerations," in *Sex Research: New Developments*, ed. John Money (New York: Holt, Rinehart and Winston, 1965), 53–89, 90–109.

13. Neiger, "Recent Trends in Sex Research," 105.

14. Colapinto, *As Nature Made Him*, 17–18.

15. Paul Rutherford, "Researching Television History: Prime-Time Canada, 1952–1967," *Archivaria* 20 (1985): 79–93, 79, 92.

16. Rutherford, "Researching Television History," 92n41.

17. Rutherford, "Researching Television History," 91.

18. Quoted in Colapinto, *As Nature Made Him*, 19.

19. For example, Robert McNair Wilson, *Pygmalion, or the Doctor of the Future* (London: Kegan Paul, 1925), 63–64.

20. John Money, *Sex Errors of the Body: Dilemmas, Education, Counseling* (Baltimore: Johns Hopkins Press, 1968), 87.

21. For example, Lawrence S. Kubie and James B. Mackie, "Critical Issues Raised by Operations for Gender Transmutation," *Journal of Nervous and Mental Disease* 147 (1968): 431–43, 431.

22. Reed Erickson, foreword to *Transsexualism and Sex Reassignment*, ed. Richard Green and John Money (Baltimore: Johns Hopkins Press, 1969), xi. This book was financed by Erickson (xvii), as was John Money and Anke A. Ehrhardt, *Man and Woman, Boy and Girl: The Differentiation and Dimorphism of Gender Identity from Conception to Maturity* (Baltimore: Johns Hopkins University Press, 1972), xiii.

23. Mission statement of the Erickson Educational Foundation, quoted in Aaron H. De-

vor and Nicholas Matte, "One Inc. and Reed Erickson: The Uneasy Collaboration of Gay and Trans Activism, 1964–2003," *GLQ* 10 (2004): 179–200, 185; Joanne Meyerowitz, *How Sex Changed: A History of Transsexuality in the United States* (Cambridge, MA: Harvard University Press, 2002), 7, 210.

24. Erickson, foreword, xi.

25. Diane Baransky, quoted in Colapinto, *As Nature Made Him*, 22.

26. Money, quoted in Colapinto, *As Nature Made Him*, 22.

27. Reimer's biographer claims that gender reassignment for Bruce had been mentioned, but not recommended, to the Reimers (Colapinto, *As Nature Made Him*, 18–19); whereas Money claims that "a consultant plastic surgeon" had recommended it (John Money, "Ablatio Penis: Normal Male Infant Sex-Reassigned as a Girl," *Archives of Sexual Behavior* 4 [1975]: 65–71, 67).

28. John Money, "Hermaphroditism: An Inquiry into the Nature of a Human Paradox" (PhD diss., Harvard University, 1952); United Nations Educational, Scientific and Cultural Organization, *The Race Concept: Results of an Inquiry* (Paris: UNESCO, 1952).

29. UNESCO, *Race Concept*, 5.

30. UNESCO, *Race Concept*, 7; Michelle Brattain, "Race, Racism, and Antiracism: UNESCO and the Politics of Presenting Science to the Postwar Public," *American Historical Review* 112 (2007): 1386–1413, 1398.

31. The responses were intended to constitute a peer review of the second statement. Unable to revise the statement in a way that synthesized the responses, the project director decided to publish them in full (Brattain, "Race, Racism, and Antiracism," 1403).

32. UNESCO, *Race Concept*, 100. This claim was influenced by the anthropologist Ashley Montagu, editor of the first statement, who had previously argued that "mankind is everywhere plastic, adaptable, and sensitive" (*Man's Most Dangerous Myth: The Fallacy of Race*, 2nd ed. [New York: Columbia University Press, 1945], 150).

33. UNESCO, *Race Concept*, 14.

34. Donna J. Haraway, *Modest_Witness@Second_Millennium.FemaleMan©_Meets_OncoMouse™: Feminism and Technoscience* (London: Routledge, 1997), 239.

35. In UNESCO, *Race Concept*, 18.

36. Sharon E. Kingsland, *Modeling Nature: Episodes in the History of Population Ecology*, 2nd ed. (Chicago: University of Chicago Press, 1995), 215–16.

37. UNESCO, *Race Concept*, 15.

38. J. W. S. Pringle, "On the Parallel between Learning and Evolution," *Behaviour* 3 (1951): 174–214, 186, 209.

39. UNESCO, *Race Concept*, 14.

40. John Money, Joan G. Hampson, and John L. Hampson, "An Examination of Some Basic Sexual Concepts: The Evidence of Human Hermaphroditism," *Bulletin of the Johns Hopkins Hospital* 97 (1955): 301–19, 308. See also John L. Hampson, Joan G. Hampson, and John Money, "The Syndrome of Gonadal Agenesis (Ovarian Agenesis) and Male Chromosomal Pattern in Girls and Women: Psychologic Studies," *Bulletin of the Johns Hopkins Hospital* 97 (1955): 207–26; John Money, Joan G. Hampson, and John L. Hampson, "Hermaphroditism: Recommendations Concerning Assignment of Sex, Change of Sex, and Psychologic Man-

agement," *Bulletin of the Johns Hopkins Hospital* 97 (1955): 284–300; John Money, Joan G. Hampson, and John L. Hampson, "Sexual Incongruities and Psychopathology: The Evidence of Human Hermaphroditism," *Bulletin of the Johns Hopkins Hospital* 98 (1956): 43–57; and two companion papers: John Money, "Hermaphroditism, Gender, and Precocity in Hyperadrenocorticism," and Joan G. Hampson, "Hermaphroditic Genital Appearance, Rearing and Eroticism in Hyperadrenocorticism," *Bulletin of the Johns Hopkins Hospital* 96 (1955): 253–64 and 265–73. Money would later state that he wrote the early Hopkins papers, even though published authorship largely was shared (John Money, *A First Person History of Pediatric Psychoendocrinology* [New York: Kluwer Academic/Plenum, 2002], 35).

41. Hampson, "Hermaphroditic Genital Appearance," 265.

42. Money, Hampson, and Hampson, "Hermaphroditism: Recommendations Concerning Assignment of Sex," 285n. Money nevertheless made several influential claims about gender's "determination," an ambivalent term that I examine in chapter 4.

43. Money, "Hermaphroditism, Gender, and Precocity in Hyperadrenocorticism," 254; Hampson, "Hermaphroditic Genital Appearance," 265; UNESCO, *Race Concept*, 100.

44. For example, Vernon A. Rosario, review of *Changing Sex: Transsexualism, Technology, and the Idea of Gender*, by Bernice L. Hausman, *Configurations* 4 (1996): 243–46, 245.

45. John Money, Joan G. Hampson, and John L. Hampson, "Imprinting and the Establishment of Gender Role," *American Medical Association Archives of Neurology and Psychiatry* 77 (1957): 333–36; Joan G. Hampson, John Money, and John L. Hampson, "Hermaphroditism: Recommendations Concerning Case Management," *Journal of Clinical Endocrinology and Metabolism* 16 (1956): 547–56.

46. Joseph H. Kiefer, "Recent Advances in the Management of the Intersex Patient," *Journal of Urology* 77 (1957): 528–36, 532.

47. Lawson Wilkins, *The Diagnosis and Treatment of Endocrine Disorders in Childhood and Adolescence*, 2nd ed. (Springfield, IL: Charles C. Thomas, 1957), 285; see also Katrina Karkazis, *Fixing Sex: Intersex, Medical Authority, and Lived Experience* (Durham, NC: Duke University Press, 2008), 60–61.

48. Lawson Wilkins, Melvin M. Grumbach, Judson J. Van Wyk, Thomas H. Shepard, and Constantine Papadatos, "Hermaphroditism: Classification, Diagnosis, Selection of Sex and Treatment," *Pediatrics* 16 (1955): 287–302, 296.

49. Wilkins, *Diagnosis and Treatment*, 209. The text attributed to Money and the Hampsons on 209–10 comes from John Money and Joan G. Hampson, "Idiopathic Sexual Precocity in the Male: Management; Report of a Case," *Psychosomatic Medicine* 17 (1955): 1–15, 14–15; and Joan G. Hampson and John Money, "Idiopathic Sexual Precocity in the Female: Report of Three Cases," *Psychosomatic Medicine* 17 (1955): 16–35, 16, 24.

50. Robert J. Stoller, Harold Garfinkel, and Alexander C. Rosen, "Psychiatric Management of Intersexed Patients," *California Medicine* 96 (1962): 30–34, 31–32; Ian Berg, Harold H. Nixon, and Robert MacMahon, "Change of Assigned Sex at Puberty," *Lancet* 282 (1963): 1216–17, 1217.

51. Money and Ehrhardt, *Man and Woman*, 119.

52. John Money and Patricia Tucker, *Sexual Signatures: On Being a Man or a Woman* (Boston: Little, Brown, 1975), 98.

53. Money, Hampson, and Hampson, "Hermaphroditism: Recommendations Concerning Assignment of Sex," 289–290.

54. John Money, "Dyslexia: A Postconference Review," in *Reading Disability: Progress and Research Needs in Dyslexia*, ed. John Money (Baltimore: Johns Hopkins University Press, 1962), 9–33, 13; Money and Ehrhardt, *Man and Woman*, 243.

55. Milton Diamond, "A Critical Evaluation of the Ontogeny of Human Sexual Behavior," *Quarterly Review of Biology* 40 (1965): 147–75, 158, 166.

56. Money, *A First Person History*, 72.

57. Kiefer, "Recent Advances in the Management of the Intersex Patient," 532; see also Berg, Nixon, and MacMahon, "Change of Assigned Sex at Puberty," 1217.

58. Colapinto, *As Nature Made Him*, 53–54.

59. Money, Hampson, and Hampson, "Hermaphroditism: Recommendations Concerning Assignment of Sex," 295; John Money, "Hermaphroditism," in *The Encyclopaedia of Sexual Behaviour*, vol. 1, ed. Albert Ellis and Albert Abarbanel (London: Heinemann, 1961), 472–84, 481 (hereafter "Hermaphroditism [encyclopedia entry]").

60. Hampson, Money, and Hampson, "Hermaphroditism: Recommendations Concerning Case Management," 551. Despite this long-standing protocol, Money claimed retrospectively that vaginal surgery was deferred in Reimer's case "to avoid postsurgical fecal infection" because "the child was not toilet-trained" (John Money, *Sin, Science, and the Sex Police: Essays on Sexology and Sexosophy* [Amherst, NY: Prometheus, 1998], 315).

61. Money, *A First Person History*, 75.

62. Money, "Hermaphroditism [encyclopedia entry]," 476.

63. Howard W. Jones Jr. and William Wallace Scott, *Hermaphroditism, Genital Anomalies and Related Endocrine Disorders* (London: Baillière, Tindall and Cox, 1958), 269; Money, "Hermaphroditism [encyclopedia entry]," 481, 477.

64. Money recommended that feminizing surgery should be done "as soon after birth as is consistent with surgical safety" ("Hermaphroditism [encyclopedia entry]," 481). For an overview of how the theory had evolved into practice by the 1990s, see Alice Domurat Dreger, *Hermaphrodites and the Medical Invention of Sex* (Cambridge, MA: Harvard University Press, 1998), 182–83. There emerged, however, disciplinary differences in gender assignment practices; when Suzanne Kessler interviewed American clinicians in 1985, she found that urologists were more inclined than endocrinologists to recommend masculinizing surgeries (*Lessons from the Intersexed* [New Brunswick, NJ: Rutgers University Press, 1998], 27–28).

65. Hampson, Money, and Hampson, "Hermaphroditism: Recommendations Concerning Case Management," 551.

66. John Money, review of *The Intersexual Disorders*, by Christopher J. Dewhurst and Ronald R. Gordon, *Journal of Nervous and Mental Disease* 152 (1971): 216–18, 218.

67. John Money, "Prenatal Hormones and Postnatal Socialization in Gender Identity Differentiation," in *Nebraska Symposium on Motivation*, ed. James K. Cole and Richard Dienstbier (Lincoln: University of Nebraska Press, 1973), 221–95, 288. Although these comments were published six years after Brenda's reassignment, Money had worked with the transsexual man in question since his first hospital registration, prior to treatment, in response to a newspaper advert announcing the opening of the Gender Identity Clinic at Johns Hopkins (Money, "Pre-

natal Hormones," 286). That announcement was made in November 1966, before Money met the Reimers (Meyerowitz, *How Sex Changed*, 219).

68. Money, "Prenatal Hormones," 286.

69. "Hermaphroditism [encyclopedia entry]," 481.

70. Meyerowitz, *How Sex Changed*, 131.

71. Carl R. Rogers, "Toward a Modern Approach to Values: The Valuing Process in the Mature Person," *Journal of Abnormal and Social Psychology* 68 (1964): 160–67, 164, 166.

72. Meyerowitz, *How Sex Changed*, 139.

73. Rogers, "Toward a Modern Approach to Values," 164.

74. Money and Ehrhardt, *Man and Woman*, 152; Carl R. Rogers, "A Note on 'The Nature of Man,'" *Journal of Counseling Psychology* 4 (1957): 199–203, 200.

75. Alfred Adler, *Understanding Human Nature* [1927], trans. Walter Béran Wolfe (London: Allen & Unwin, 1928), 18.

76. Adler, *Understanding Human Nature*, 62, 35.

77. John Money, *The Psychologic Study of Man* (Springfield, IL: Charles C. Thomas, 1957), 51; Adler, *Understanding Human Nature*, 23.

78. Adler, *Understanding Human Nature*, 6.

79. Adler, *Understanding Human Nature*, 23; Money, *Psychologic Study of Man*, 48.

80. Money, "Psychosexual Differentiation," 12.

81. Money, *Psychologic Study of Man*, 54; John Money and Clay Primrose, "Sexual Dimorphism and Dissociation in the Psychology of Male Transsexuals," *Journal of Nervous and Mental Disease* 147 (1968): 472–86, 481.

82. Elizabeth Haiken, *Venus Envy: A History of Cosmetic Surgery* (Baltimore: Johns Hopkins University Press, 1997), 112–14.

83. Adler, *Understanding Human Nature*, 35.

84. John Money, "Unanimity in the Social Sciences with Reference to Epistemology, Ontology, and Scientific Method," *Psychiatry* 12 (1949): 211–21, 220.

85. Adler, *Understanding Human Nature*, 74.

86. Adler, *Understanding Human Nature*, 41, 69.

87. B. K. Rank, "The Story of Plastic Surgery, 1868–1968," *Practitioner* 201 (1968): 114–21, 115–17.

88. Haiken, *Venus Envy*, 108–9, 115.

89. Jacques W. Maliniak, *Sculpture in the Living: Rebuilding the Face and Form by Plastic Surgery* (New York: Romaine Pierson, 1934), 196.

90. Maliniak, *Sculpture in the Living*, 196, caption to figure 22 opposite p. 65.

91. Maliniak, *Sculpture in the Living*, 76.

92. Maliniak, *Sculpture in the Living*, 87, 138.

93. Maliniak, *Sculpture in the Living*, 77, 138.

94. Jacques W. Maliniak, "Your Child's Face and Future," *Hygeia* (May 1935): 410–13, 411.

95. Maxwell Maltz, *Doctor Pygmalion: The Autobiography of a Plastic Surgeon* (London: Museum Press, 1954), 174; see Bernice L. Hausman, *Changing Sex: Transsexualism, Technology, and the Idea of Gender* (Durham, NC: Duke University Press, 1995), 53–54.

96. Money, "Prenatal Hormones," 256; see also Money and Ehrhardt, *Man and Woman*, 152.

97. Hampson, Money, and Hampson, "Hermaphroditism: Recommendations Concerning Case Management," 549. In the 1940s, some clinicians included among the effects of intersex "feelings of social-sexual inferiority," but did not present this as curable, merely as a diagnosis (Albert Ellis, "The Sexual Psychology of Human Hermaphrodites," *Psychosomatic Medicine* 7 [1945]: 108–25, 119; Jacob E. Finesinger, Joe V. Meigs, and Hirsh W. Sulkowitch, "Clinical, Psychiatric and Psychoanalytic Study of a Case of Male Pseudohermaphroditism," *American Journal of Obstetrics and Gynecology* 44 [1942]: 310–17, 315). The explicit use of the term "inferiority" in medical papers on intersex continued occasionally until at least the mid-1970s, including by authors who disagreed with Money over aspects of treatment (Daniel Cappon, Calvin Ezrin, and Patrick Lynes, "Psychosexual Identification [Psychogender] in the Intersexed," *Canadian Psychiatric Association Journal* 4 [1959]: 90–106, 98; Berg, Nixon, and MacMahon, "Change of Assigned Sex at Puberty," 1217; Arye Lev-Ran, "Gender Role Differentiation in Hermaphrodites," *Archives of Sexual Behavior* 3 [1974]: 391–424, 405). Money never directly called intersex a problem of inferiority, even though in 1961 he cited the aforementioned 1940s papers (John Money, "Sex Hormones and Other Variables in Human Eroticism," in *Sex and Internal Secretions*, 3rd ed., vol. 2, ed. William C. Young [Baltimore: Williams and Wilkins, 1961], 1383–1400, 1384). I argue that inferiority was nonetheless the "problem" that the surgical and social construction of "gender" was implied by Money to solve.

98. Money, *A First Person History*, 73; Money, *Sin, Science, and the Sex Police*, 316.

99. Money and Ehrhardt, *Man and Woman*, 120. Money also stated incorrectly that the circumcision took place at age seven months, not eight months (118), an error repeated in his other reports on the case (for example, "Prenatal Hormones," 293, caption to figure 59).

100. "Biological Imperatives," *Time*, January 8, 1973, 34.

101. Money, "Prenatal Hormones," 226.

102. Money, "Prenatal Hormones," 292.

103. Money, "Prenatal Hormones," 293, 294. On the wider context, see Rebecca M. Jordan-Young, *Brain Storm: The Flaws in the Science of Sex Differences* (Cambridge, MA: Harvard University Press, 2010), 36.

104. Money, "Prenatal Hormones," 294; Money and Tucker, *Sexual Signatures*, 97, 98.

105. Money, *A First Person History*, 76.

106. Money and Ehrhardt, *Man and Woman*, 144–45, 18.

107. Money and Tucker, *Sexual Signatures*, 91, 98.

108. Money, *Sin, Science, and the Sex Police*, 318.

109. Money, "Ablatio Penis," 65.

110. John Money, "Nativism versus Culturalism in Gender-Identity Differentiation," in *Sexuality and Psychoanalysis*, ed. Edward T. Adelson (New York: Brunner/Mazel, 1975), 48–64, 57.

111. Money, "Ablatio Penis," 65.

112. Money and Ehrhardt, *Man and Woman*, 119; Money, "Ablatio Penis," 67; Money, *Sin, Science, and the Sex Police*, 315.

113. Money and Tucker, *Sexual Signatures*, 92–93; Money, *A First Person History*, 73; Colapinto, *As Nature Made Him*, 49–50.

114. John Money and Mark Schwartz, "Biosocial Determinants of Gender Identity Dif-

ferentiation and Development," in *Biological Determinants of Sexual Behaviour*, ed. John B. Hutchison (Chichester: John Wiley, 1978), 765–84, 779.

115. Money and Schwartz, "Biosocial Determinants," 780.

116. Kessler, *Lessons from the Intersexed*, 6.

117. David Haig, "The Inexorable Rise of Gender and the Decline of Sex: Social Change in Academic Titles, 1945–2001," *Archives of Sexual Behavior* 33 (2004): 87–96, 93–94.

118. For example, Carol Tavris and Carole Offir, *The Longest War: Sex Differences in Perspective* (New York: Harcourt Brace Javonovich, 1977), 109–10; Rhoda K. Unger, *Female and Male: Psychological Perspectives* (New York: Harper and Row, 1979), 134–36.

119. Shirley Weitz, *Sex Roles: Biological, Psychological, and Social Foundations* (New York: Oxford University Press, 1977), 56 (see also 51); Ian Robertson, *Sociology*, 2nd ed. (New York: Worth, 1981), 344–45 (see also 314).

120. JoAnn Brooks, review of *Sexual Signatures: On Being a Man or a Woman*, by John Money and Patricia Tucker, *Library Journal*, April 1, 1975, 677; Charles C. Gross, review of *Sexual Signatures: On Being a Man or a Woman*, by John Money and Patricia Tucker, *Signs* 1 (1976): 742–44, 742.

121. Critiques included Barbara Fried, "Boys Will Be Boys Will Be Boys: The Language of Sex and Gender," in *Women Look at Biology Looking at Women: A Collection of Feminist Critiques*, ed. Ruth Hubbard, Mary Sue Henifin, and Barbara Fried (Boston: G. K. Hall, 1979), 39–59, 48; and Alexandra G. Kaplan, "Human Sex-Hormone Abnormalities Viewed from an Androgynous Perspective: A Reconsideration of the Work of John Money," in *The Psychobiology of Sex Differences and Sex Roles*, ed. Jacquelynne E. Parsons (Washington, DC: Hemisphere, 1980), 81–91, 85.

122. Lesley Rogers, "The Ideology of Medicine," in Dialectics of Biology Group, *Against Biological Determinism*, ed. Steven Rose (London: Allison and Busby, 1982), 79–93, 88.

123. Quoted in Milton Diamond, "Sexual Identity, Monozygotic Twins Reared in Discordant Sex Roles and a BBC Follow-Up," *Archives of Sexual Behavior* 11 (1982): 181–86, 184.

124. Diamond, "Sexual Identity," 183.

125. Money, *Sin, Science, and the Sex Police*, 316–17, 317.

126. Diamond, "A Critical Evaluation," 150. Money and his colleagues also argued for prenatal hormonal influences on gender roles, but distinguished this from gender identity, which they claimed to be determined by rearing (Susan W. Baker, "Biological Influences on Human Sex and Gender," *Signs* 6 [1980]: 80–96, 83). Nevertheless, Money criticized others for separating gender identity from role (*A First Person History*, 36).

127. Diamond, "Sexual identity," 184.

128. Colapinto, *As Nature Made Him*, 207–9.

129. Milton Diamond and H. Keith Sigmundson, "Sex Reassignment at Birth: Long-Term Review and Clinical Implications," *Archives of Pediatric and Adolescent Medicine* 151 (1997): 298–304, 299–300, 302.

130. John Colapinto, "The True Story of John/Joan," *Rolling Stone*, December 11, 1997, 54–73 and 92–97, 56, 95; Colapinto, *As Nature Made Him*, xiii–xv.

131. Milton Diamond and H. Keith Sigmundson, "Management of Intersexuality: Guidelines for Dealing with Persons with Ambiguous Genitalia," *Archives of Pediatric and Adolescent*

*Medicine* 151 (1997): 1046–50, 1047–48; Milton Diamond, "Pediatric Management of Ambiguous and Traumatized Genitalia," *Journal of Urology* 162 (1999): 1021–28, 1025.

132. Diamond and Sigmundson, "Management of Intersexuality," 1047.

133. Diamond and Sigmundson, "Management of Intersexuality," 1050.

134. Hampson, "Hermaphroditic Genital Appearance," 266.

135. Money also briefly alleged in 1991 that Diamond's "alliance" with the BBC had "prematurely terminated" the case, but he did not engage with the BBC's claims or Diamond's 1982 article (John Money, *Biographies of Gender and Hermaphroditism in Paired Comparison: Clinical Supplement to the Handbook of Sexology* [Amsterdam: Elsevier, 1991], 10).

136. Money, *Sin, Science, and the Sex Police*, 318.

137. Money, *Sin, Science, and the Sex Police*, 318–19.

138. Money and Tucker, *Sexual Signatures*, 97.

139. Money, *Sin, Science, and the Sex Police*, 319.

140. Colapinto, *As Nature Made Him*, 186.

141. Money, *A First Person History*, 75; Colapinto explained this error in "The Boy without a Penis," *The Position*, accessed March 9, 2013, http://web.archive.org/web/200012180124/http://www.theposition.com/takingpositions/usconfidential/00/09/18/colapinto/default.shtm, para. 19 of 22.

142. Unlike most of the book in which it appeared, the 1998 chapter had not previously been published in a journal or delivered at a conference (Money, *Sin, Science, and the Sex Police*, 297).

143. Richard Green, preface to *A First Person History of Pediatric Psychoendocrinology*, by John Money (New York: Kluwer Academic/Plenum, 2002), vii–viii, vii.

144. Richard Green, "John Money, Ph.D. (July 8, 1921–July 7, 2006): A Personal Obituary," *Archives of Sexual Behavior* 35 (2006): 629–32, 630. Green refers specifically to his experience on the editorial board of the *Archives of Sexual Behavior*; in correspondence with me on April 18, 2011, Green stated that he did not know whether Money's practice extended to his submissions elsewhere. (I wrote to Anke Ehrhardt on June 2, 2011, to ask whether she could comment on the generalizability of Green's remark; I did not receive a reply.) I think it is exceedingly unlikely that Money would have adopted such an approach to one journal, but not to other publications. As his colleague Vern Bullough recounted, "Money was a hard man to convince to change once he made up his mind. I tried to get him to change the title of *Venuses Penuses*, a book for which I was general editor, but he was determined to keep the title. . . . One is advised not to argue with John" ("The Contributions of John Money: A Personal View," *Journal of Sex Research* 40 [2003]: 230–36, 235).

145. Money, *Sin, Science, and the Sex Police*, 297, 302; 304–9; Money, *A First Person History*, 71.

146. Georgia Warnke, "The Tragedy of David Reimer," in *After Identity: Rethinking Race, Sex, and Gender* (Cambridge: Cambridge University Press, 2007), 15–48.

147. Intersex Society of North America, "Who Was David Reimer (Also, Sadly, Known as 'John/Joan')?," accessed March 9, 2013, http://www.isna.org/faq/reimer, para. 4 of 5.

* 2 *

# *Vandalizing*

CHAPTER 4

# Cybernetic Sexology

*Iain Morland*

John Money claimed to think "cybernetically" about sex, gender, and sexuality.[1] Using a cybernetic vocabulary of "variables," "thresholds," and "feedback systems," Money purported to offer a more scientific and up-to-date sexology than hitherto possible.[2] In this chapter, I will present the first-ever critical analysis of Money as a cybernetic theorist. Evaluating the context and rhetoric of his claims, I will explain how Money used cybernetics—the study of communication and control, conceived during 1940s military research—to distance sexology from both psychoanalytic and biological studies. Further, I will show how cybernetic theory shaped Money's treatment recommendations for individuals with ambiguous genitalia. However, I will also critique a formative error made by Money in his application of cybernetics to sexology. Cybernetics theorized dynamic systems that can adapt, not merely repeat. It was therefore irreconcilable with the sudden, irrevocable establishment of gender in infancy that was axiomatic for Money. Consequently, I will argue that Money's model of gender, as reiterated without conscious reflection, was closer to a psychoanalytic view of unconscious determinism than a cybernetic theory.

## MONEY'S CYBERNETIC CONTEXT

Recently, Money's invention of the nomenclature "gender role" has been interpreted by cultural critic Jennifer Germon as evidence that sociology shaped Money's thought. The present chapter has a similar starting point, but traces a different genealogy. Germon argues that Money's uptake of "role" reflects the ideas of Talcott Parsons, the influential sociologist whose seminars Money attended at Harvard. Parsons analyzed social institutions, including roles, in terms of their function in current society. Explained functionally, social institutions are like bodily organs, for they "perform specific tasks and contribute toward the maintenance of a homeostasis or equilibrium," as Germon puts it.[3] She notes that this is an inherently conservative approach to prevailing social arrangements, and even a type of "circular reasoning" whereby, for example, the effects of a role are interpreted as its purpose.[4] I agree that homeostasis was important to Money, but I think that Germon overstates sociology's influence on his thought. She overlooks the contemporary discipline for which "circularity of action" was not an error of reasoning but an object of study, and for which homeostasis was a signature concern: cybernetics.[5]

## Defining the Disciplines

In a little-known 1949 paper about the social sciences, Money cited approvingly an article from a recent issue of *Scientific American*.[6] Billed by the magazine as presenting "an intriguing new approach to the comparative study of men, machines and societies," the eponymous subject of Money's source was cybernetics.[7] Its author, American mathematician Norbert Wiener, created the term "cybernetics" in 1947. His neologism referred to the design and analysis of systems that behave according to information transmitted between their constituent parts.[8] Wiener had developed this idea during World War II, when he researched improvements to the accuracy of his country's antiaircraft artillery. Wiener sought to increase the quality of the information that determined the direction of fire. Treating radar information about an aircraft's trajectory as feedback to control a gun's direction, he designed a weapon that regulated its motion in anticipation of its target's next position.[9] Wiener's artillery research in the early 1940s transpired to have little practical impact.[10] However, it was foundationally important to cybernetic theory, because it enabled Wiener to conceptualize guns and pilots as functionally equivalent: both relied upon feedback to achieve their respective goals of shooting down aircraft and dodging gunfire. Moreover, the pilot's evasive maneuvers constituted the very feedback used by the gun, so they were really a single large system.[11] As I will argue, such an expansive understanding of what Wiener called "communication and control" systems simultaneously helped and hindered the dissemination of cybernetics, and structured Money's corresponding efforts to delineate sexology and its nomenclature.[12]

Homeostasis was central to cybernetics because it described the behavior of successfully self-regulating systems. Like cybernetics, homeostasis has a precise origin: coined in 1926 by physician Walter Cannon, the word referred initially to the concurrent openness and resilience of living organisms to their surroundings. "Changes in the surroundings excite reactions," wrote Cannon, but "automatic adjustments" mean that "wide oscillations are prevented and the internal conditions are held fairly constant."[13] The regulation of body temperature is an example. Cybernetic theorists expanded the definition of homeostasis beyond organic systems.[14] Wiener's antiaircraft gun, for instance, could be called homeostatic when hitting a moving target multiple times. It would attain that state not by remaining still, but by adapting to follow the target, just as body temperature is maintained through continuous adjustments. Further, if the gun were to hit two targets in succession, it would enter what cyberneticists called an "ultrastable" state, successfully maintaining homeostasis across a major change in the system, the destruction of one aircraft and the appearance of another.[15] Indeed, Cannon had stressed that homeostasis does not mean immobility or "stagnation," but rather a dynamically produced "steady state."[16] Nevertheless, the ambivalence of the homeostasis concept—stability and adaptation together—would inform Money's mishandling of cybernetics, as we will see.

The generalization of homeostasis in cybernetic theory, to describe systems that were structurally diverse but functionally homologous, meant that cybernetics from its inception was irreducibly interdisciplinary. In his *Scientific American* article, Wiener named no less than eleven disciplines represented by his coworkers, including anthropology, engineering, physics, and psychology.[17] The reason for the mix of specialisms, Wiener explained, was that cybernetics seeks "common elements in the functioning of automatic machines and of the human nervous system."[18] The antiaircraft gun and pilot, each learning from the other's behavior, were the wartime template for these commonalities. Wiener envisaged civilian applications for cybernetics too, such as "the design of control mechanisms for artificial limbs"—perhaps intended, in a further instance of how cybernetics connected disciplines, for airmen injured by artillery.[19] The academic identity of cybernetics, then, was shaped by its homologies: one contemporary reviewer of Wiener's work commented that its classification as "literature or science" hinged on the particulars of how "machines closely resemble men in their activities."[20] Money certainly thought there was a resemblance: in 1956, he likened the brain to a telephone exchange—a homology lifted from Wiener's *Scientific American* article.[21] In short, cybernetics provided Money, in his career's formative years, with a model for the relationship between interdisciplinarity and scientific credibility.

As Money went on to articulate his vision for sexology, he expressed ambitions and anxieties about interdisciplinarity that mirrored those of the cyberneticists. Just as Wiener had argued that "the most fruitful areas for the growth of the sciences" are the "no-man's lands between the various established fields," requiring interdisciplinary teams for their development, so too did Money propose in 1973 that fields of sex research should not be kept "separated as though unrelated," but rather unified into "an organized body of sexological theory."[22] The benefit from such a sexology, in Money's view, would be the prevention of disputes over "nature vs. nurture" and "soma vs. psyche"—a claim that echoed Wiener's description of cybernetics as "not strictly biological or strictly physical, but a combination of the two."[23] But Money and the cybernetic theorists, such as psychiatrist Ross Ashby, also insisted that their respective fields were not derivative amalgamations of current disciplines. For Money, sexology's grounding in interpersonal relations enabled it to tackle problems unsolvable by "individual traditions of treatment" like gynecology and endocrinology; for Ashby, cybernetics not only gave a "common language" to existing scientific traditions, but revealed truths independently, for it possessed "its own foundations."[24] Likewise, what sexology needed, Money claimed, was a language as scientifically universal as mathematics.[25]

I wish to argue, then, that Money's coining of "gender" in 1955 paralleled Wiener's "cybernetics" from eight years earlier. Both were attempts to overcome the difficulties of interdisciplinarity, using language to assert scientific authority. I recognize that Money's sexological project might appear very different from cybernetics, despite the connections that I have identified so far. The twentieth-century humanism wherein "gender" was forged, and which I discussed in chapter 3, may look irreconcilable with the mechanistic interests of cybernetics, exemplified by homeostasis and military research. However, Cannon and Wiener positioned their work as humanist. Cannon, analyzing functional analogies between "the physiological and social realms," distinguished the inner constancy of physiological homeostasis from the internal dynamics of social organizations.[26] The cause of the difference was the human brain—"an organ for discovery and invention," driving social improvement, creating disruption, and enabling adaptation to change.[27] Correspondingly, Wiener distanced cybernetics from the "fascist" ideal of a rigidly structured nation wherein individuals would perform repetitive, highly specialized functions.[28] He argued for humanity's organically inherent "variety and possibility," and stated in *Scientific American* that the neonatal nervous system showed especially "great flexibility."[29] A similar belief in human plasticity, as

I chronicled in chapter 3, underpinned Money's treatment recommendations for intersex.

So, for Money in the mid-1950s, cybernetics was a topical precedent in the creative use of language to define a discipline. Already in 1949, he had recommended that researchers from overlapping specialisms should communicate with a shared vocabulary, in the same way that Wiener sought to overcome "a lack of unity" in his field by coining "cybernetics" from the Greek for "steersman," representing communication and control in self-regulating systems.[30] Compellingly for Money, cybernetics theorized the adaptive consequences of plasticity: it combined the study of homologous adaptations in organisms and machines with a humanist discourse about the exceptional plasticity of people. I propose, therefore, that Money did not merely borrow cybernetic rhetoric in articulating gender. For Money, gender was cybernetic, directly. His innovative use of the term—for "a person's conviction of himself as a man or herself as a woman"—was a claim about the emergence of homeostasis from infantile plasticity.[31] It was also indivisible from Money's aspirations for sexology, an effort to show that "sexology belongs to science" through the universality of its nomenclature.[32] But such a strategy, I will reveal, tethered sexology to an idea of communication and control, with unexpected ramifications for intersex treatment.

## From Causality to Prediction

I am going to argue that gender, like cybernetics, described the convergence of communication and control in the prediction of phenomena. Moreover, a focus on prediction rather than causality meant that Money shared with cyberneticists another strategy for asserting scientific authority, the claim to be up to date—to espouse "What Is New," as the opening chapter of Ashby's *Introduction to Cybernetics* was titled crisply.[33] Professing to avoid causal explanations of human behavior, Money objected that such explanations evoked "platitudes of dichotomy," which were "worn-out." Despite his association with the question of "nature versus nurture," it was among several dichotomies that Money criticized as redundant, as Lisa Downing noted in chapter 2. They included biological versus social, physiological versus psychological, and most relevantly to cybernetics, heredity versus environment.[34] In cybernetic scholarship, attention to homeostasis had rendered obsolete the separation of heredity and environment, replaced by the analysis of their interaction.[35] Accordingly, Money stated that although genes were commonly considered an ineradicable inheritance, they were alterable by "environmental interventions."[36] Likewise, to seek purely hereditary or environmental causes for phenomena ranging from homosexuality to sexual inequality was "outdated" and "outmoded," he insisted.[37] Money cast cybernetics, contrastingly, as scientifically progressive, inspired by Wiener's narrative of succession from dichotomous thought to the understanding of "circular processes," which Wiener called "a step forward."[38]

In sexology and cybernetics alike, the claim to be up to date was more complex and ambivalent than it might seem. It was complex because it entailed distinguishing present-day science from views that were historically prior, and also from contemporary views that were allegedly defunct. For example, Ashby differentiated cybernetic theory not simply from assumptions made in "mediaeval and earlier times," but from current ideas based on those assumptions.[39] The claim to be up to date was ambivalent because even while characterizing dichotomous thought as obsolete, it invoked a dichotomy between up to date and out of date. Distinguishing his work from earlier views, Money asserted that all redundant dichotomies built on a "very ancient" belief in the separation of body and mind. Incongruously, he used dichotomously periodizing terms to describe this belief as "pre-Platonic" and even "pre-Biblical."[40] The separation of body and mind was inapplicable to sexology, Money declared, because sex exists equally "in the head" and "in the groin."[41] However, he was concerned that others disagreed. Like Ashby, Money criticized the persistence of old beliefs in the present, complaining that the body versus mind dichotomy was entrenched in "folk metaphysics." Money urged scientists to overcome the dichotomy, in order to avert "disaster" in sexology—an exhortation that entangled the up to date and the out of date, by dichotomizing science versus nonscience.[42]

For Money and the cyberneticists, criticizing specific dichotomies served the deeper purpose of debunking the distinction between cause and effect. They wanted to show that behavior was not explicable by the attribution of causality to either side of any dichotomy.[43] This critique was especially relevant to psychology, in Money's opinion, because in psychology, behavior was theorized typically as either unconsciously or biologically motivated. Such theories were not really opposed, Money claimed: both interpreted behavior as causally determined. Further, if behavior had a deterministic explanation, then it was involuntary. While Money rejected psychology's assumption of causality, he agreed about the involuntariness of behavior.[44] Turning to cybernetics to articulate his position, Money replaced the psychological concepts of motivation and drive with what he called "thresholds of behavioral release."[45] This characteristically cybernetic phrase described behavior in self-regulating systems. It referred to a system's regulation of its behavior in response to changes in its own state—for example, in an organism, sweating when body temperature exceeds a threshold; or in an antiaircraft gun, firing when a target's distance is below a threshold. Crossing a threshold releases a behavior. The concept of the threshold could explain how part of a system changed only "when the disturbance coming to it exceeds some definite value," Ashby wrote.[46]

Money's assertion that thresholds of behavioral release operate involuntarily is demonstrated by the homeostatic maintenance of body temperature in a steady state.[47] His point about the noncausality of such mechanisms is less clear. One might regard the crossing of a threshold as a cause and behavior as its effect. However, that would repeat the reasoning that Money rejected. He lamented that "whenever men speak of two things, then they will surely, sooner or later, allow one to be the cause, the other effect."[48] To understand the concept of the threshold cybernetically, and to clarify its significance for Money, threshold and behavior should be grasped as one "circular process," in Wiener's phrase. The process is circular because threshold and behavior exist within a single system, enabling self-regulation through the action—or inhibition of action—of one part on another part. To self-regulate, a system must circulate information about its current state relative to the threshold: the body sweats, but not when it has already cooled; the gun fires, but not after the aircraft has been destroyed. Feedback, then, closes the loop between threshold and behavior. For Money, the mechanism for feedback was the mind itself. Calling the mind "the information and communication function of the human organism," Money placed cybernetics at the center of psychology.[49]

Having eschewed causality, Money and the cyberneticists oriented their respective disciplines around the prediction of behavior, rather than speculation about its causes. Ashby described cybernetics as the study of all "regular," "determinate," or "reproducible" behaviors; similarly for Money, sexology was a quest for "recurrent temporal sequences" regarding sex, within his declared wider project of cataloging behavior.[50] The common orientation of the disciplines was significant, because irrespective of how abstruse the rejection of causality might have sounded to critics, the examination of behavioral patterns gave cybernetics and sexology an apparently conventional research agenda. Money even stated that the viability of any science, not just sexology, depended on its "predictive power."[51] But in shifting from causality to prediction, cybernetics and sexology broke from orthodox criteria for scientific accuracy. In place of accurate reference to a presumed external reality, they sought accurate predictions about self-contained systems. So, for instance, whereas accuracy of reference could reveal the current position of an aircraft, only accuracy of prediction could specify its position in the future.[52] Indeed, Wiener termed antiaircraft systems "prediction machines."[53] Replacing causality with prediction meant replacing the pursuit of knowledge about the present with the pursuit of certainty about the future. I am going to argue that gender was sexology's own prediction machine.

Money coined "gender" to describe sexual development in a noncausal, but nevertheless controllable, way. This formulation was integral to his clinical influence. For doctors and parents to whom sexually ambiguous infants were disquieting, Money's theory of gender provided reassurance. Ostensibly, it enabled the prediction of an infant's future self-perception as female or male, untethered from the effects of sexual ambiguity. The predictability of gender was nothing less than the justification for treatment.[54] Money claimed that his treatment recommendations, which prioritized unambiguous rearing as either female or male, made gender in intersex infants "to all intents and purposes predictable."[55] Comparatively, Money asserted, chromosomes, gonads, and hormones were not reliable predictors of gender, let alone its cause; and out-of-date psychological theories failed similarly to offer "prediction and control" over sexual development.[56] Accordingly, Money interpreted "those instances where the prediction falls down" as failures of communication and control. If "uncertainty as to the sex of the baby at birth" were "transmitted" to parents, they would, like an errant telephone exchange, "transmit mixed signals to the child"; and if reassignment occurred too late in life, a patient would be "not predictably free" from psychological problems.[57] Successful treatment, then, did not require accurate reference to an infant's anatomy—for awareness of sexual ambiguity could even be detrimental—but, rather, certainty in the prediction of gender. It meant specifying in advance the steady state of gender that would follow from the plasticity of infancy. Treatment, by Money's definition, was a cybernetic accomplishment.

## INSIDE THE PREDICTION MACHINE

I am suggesting that gender, in Money's cybernetic sexology, was a prediction machine for the communication and control of sexual development. In the previous section, I uncovered the cybernetic context of Money's work, examining not only the parallels between sexology and cybernetics, but also their convergence in Money's formulation of gender. I introduced three key cybernetic concepts—homeostasis, feedback, and threshold. The second of these, feedback, will be explored in greater detail in the current section; I will discuss the threshold in the section that follows. Readers may be concerned, though, that my trajectory will be simply to criticize the concepts in turn. That is not the case. I want to tread between some recent positions on Money's work. Whereas Germon has suggested that gender should be reappraised to restore "the dignity and humanity of those who have historically been most marginalized by it," the critic David Rubin has countered that the "technologies of psychosomatic normalization" mobilized by gender are more intractable in their "regulatory power."[58] Following my critique of humanism in chapter 3, I am skeptical about Germon's proposal, and I agree broadly with Rubin. However, I propose that by looking inside the prediction machine called "gender," we can discover how Money, despite his uptake of cybernetics, mishandled the theory on which the communication and control of sexual development might have relied. My critique, then, will not be that Money successfully wielded cybernetics to regulate sexual development. In fact, he failed to think truly cybernetically. Money's error centered, I will argue, on the meaning of feedback.

## The Variables of Sex

The prediction machine appeared to work by partitioning an infant's sex into discrete elements, including physical components (such as chromosomes and hormones) and social ones (such as sex of rearing), then using a certain constellation of these elements to predict the infant's future gender. The portentous elements were suggested by Money to exist in feedback relations, acting upon one another. Before I scrutinize this mechanism and its elements, the very fact that Money partitioned sex requires analysis. It indicates a tension between Money's clinical engagement with intersex, his theoretical stance against dichotomous thought, and his efforts to define sexology. "In my earliest studies of hermaphroditism," Money recalled during the 1980s, "I came to the realization that there is no absolute dichotomy of male and female. A person's sex must be specified not on the basis of a single criterion, but of multiple criteria."[59] Although this comment seems logical, the second sentence is a non sequitur. The first sentence implies that a person's sex cannot be "specified" at all—a conclusion that would have resonated with Money's professed rejection of dichotomous thought. However, in defining sexology, Money retained the dichotomy of male versus female. For him, sexology was the science of "the differentiation and dimorphism of sex."[60] In other words, the dichotomy of sexual dimorphism was central to Money's scientific project. The non sequitur reflects this: Money followed the observation of "no absolute dichotomy of male and female" with a proliferation of dichotomies, rather than a recognition of their redundancy. Dichotomies proliferated among the "multiple criteria" of sex, each of which could be judged in most cases as male or female—male chromosomes, female hormones, and so on. This move enabled Money to disambiguate sexual ambiguity, explaining it in terms of "contradictions" between elements that, taken singly, were typically dichotomously sexed.[61] It was a shift from what Money termed a "unitary" definition of sex to a "multivariate" one—not postdichotomous, but specified by multiple variables.[62]

Seven variables of sex composed Money's prediction machine. First presented in a 1955 paper, the variables were "chromosomal sex; gonadal sex; hormonal sex; external genital morphology; accessory internal genital morphology; assigned sex and rearing; and gender role."[63] In further publications, Money reiterated the list with small changes: later in the 1950s, he grouped "hormonal sex" with "secondary sexual characteristics," which he rephrased subsequently as "pubertal feminization or virilization."[64] But in 1973, in the light of research that purported to show gendered consequences of prenatal exposure to hormones, Money reformulated the variable as "Hormonal sex: (a) prenatal and (b) pubertal."[65] The only other variable that changed was gender itself. It was first called a "variable" by Money in April 1955, at which time he articulated it as "gender role and erotic orientation," a fortnight before formulating the full list of seven variables.[66] (For this reason, I regard it as a metonymic name for the entire prediction machine.) In later publications, gender appeared as "gender role, including psychosexual orientation" and, foregrounding its sexually dichotomous character, "gender role and orientation as male or female."[67] It was then disaggregated by Money in 1965 into "gender role and identity," before coalescing eight years later into "gender identity/role."[68] Throughout these transformations, Money treated gender as a single variable. Occasionally he bracketed gender with "assigned sex and rearing," distinguishing those two variables from the five "physical" ones.[69] More frequently, Money separated only gender, supposedly to examine its relation to each of the six other variables.[70]

The multivariate definition of sex was a mark of what Money called the "scientific way" of understanding phenomena—that is, identifying "some of the contributing variables," then assessing "the relative importance of each."[71] It was also the cybernetic way. In cybernetic theory, systems were conceptualized in terms of variables selected for study. According to Ashby, objects in the world contained unlimited variables, so the task of cybernetics was not to examine every variable, but to "pick out and study the facts that are relevant to some main interest that is already given."[72] Despite its pedestrian appearance, this method had a profound consequence: in cybernetics, a system was "not a thing, but a list of variables" chosen by an observer.[73] For example, in the system that regulates body temperature, an observer might even include the variable position of the earth relative to the sun. The definition of systems as sets of observer-selected variables allowed cyberneticists to recognize a world of boundless complexity, but also to constrain it, much like Money's multivariate definition of sex. Hence, both Money and Ashby stressed that variables had to be rigorously discrete. The former insisted that any classificatory criteria, in order to be scientific, needed to operate "without overlap or ambiguity"; while the latter defined a variable as "*a measurable quantity which at every instant has a definite numerical value*."[74] Notwithstanding the kindred tone of these statements, Ashby's mathematical nomenclature could appear incompatible with Money's sexological territory. However, in a novel instance of cybernetics borrowing from sexology to exemplify a point, Ashby referenced sexual dimorphism, invoking the unit of the bit—"a contraction of 'BInary digiT'"—to assert that "the variety of the sexes is 1 bit."[75] In this respect, the dichotomous understanding of sex in sexology was congruent with the numerical understanding of variables in cybernetics.

## The Ambivalence of Determination

The discreteness of the seven sexual variables, although important to the scientific credibility of the system that comprised them, led Money to two contradictory positions. On the one hand, he stated that the variables were not simply distinct, but autonomous. Starting from the relatively circumspect claim that gender was "a variable quite independent of genes and chromosomes," Money went on to propose that the seven variables "may vary independently of one another."[76] He claimed to have learned about the independence of the variables from studying intersex individuals, because in "ordinary, healthy people," they existed in "perfect correlation"—a word choice that eschewed causality between even concordant variables.[77] So ordinary was the concordance of variables that "most people," Money declared, react with "horror" upon discovering a contradiction between sex variables in themselves or their children, mistakenly regarding it as a cause of homosexuality.[78] A calmly scientific recognition that contradictions between variables were possible, because the variables were independent, was one position that Money adopted. However, on the other hand, he argued against the independence of the variables. The reason was that in order for gender to be predictable, Money needed to posit a relation between gender and at least one other variable, which could be used for its prediction. The predictive variable, according to Money, was "sex of assignment and rearing."[79]

I argue that the contradiction between Money's positions is signaled by the ambivalence of the term "determination" in his writings, along with related words such as "determinant," for example in the 1978 paper title "Biosocial Determinants of Gender Identity Differentiation and Development."[80] Determination could describe the control of gender, but could also describe its prediction. Hence, Money moved between saying that gender was not "automatically or instinctively determined" by any of the five physical variables (chromosomes, gonads, hormones, external genitals, internal genitals) and saying that such variables failed as "indices which may be used to predict" gender.[81] This ambivalence was a misstep, I think, in Money's progress from causality to prediction, because it could be read causally and predictively, all at once. Money judged each physical variable to be neither a "direct, automatic determinant" of gender (causally) nor a reliable "prognosticator" of it (predictively), and never clarified whether the failure of prediction was a result of the absence of causality or a separate matter.[82] Nonetheless, this was a productive misstep for Money, because it allowed his treatment recommendations about the prediction of gender to be interpreted as directives for gender's control. So, when Money named sex of assignment and rearing as "consistently and conspicuously a more reliable prognosticator" of gender than any other variable, he had already provided a way to read prediction as causality.[83] In other words, the prediction of gender, on the basis of assigned sex and rearing, could mean the control of gender *by* assigned sex and rearing. Such complex ambivalence constituted the "determination" of gender.[84]

The ambivalence of gender's determination is traceable to another major scientific influence on Money's thought. It originates with the American physicist Percy Bridgman. In the 1940s, Money read Bridgman's 1927 book *The Logic of Modern Physics*, which he would later describe as having "influenced every part of my research life, all of my writings, and I think it correct to say, my entire approach to life."[85] Despite this arresting claim, the impact of Bridgman on Money has received no critical attention. During Money's graduate studies at Harvard, he attended seminars by Bridgman that arguably were more formative than those he took with Parsons.[86] Money was especially interested in Bridgman's operationism. In the same paper where he cited Wiener, Money glossed operationism as the theory that some phenomena "can be described and defined only in terms of the operations whereby one becomes aware of them."[87] Concatenating the processes of discovery and definition, operationism taught Money the ambivalence of determination before he began publishing on gender. Bridgman's use of the word "determined" in *The Logic of Modern Physics* was instructive on this point:

> To find the length of an object, we have to perform certain physical operations. The concept of length is therefore fixed when the operations by which length is measured are fixed: that is, the concept of length involves as much as and nothing more than the set of operations by which length is determined. In general, we mean by any concept nothing more than a set of operations; *the concept is synonymous with the corresponding set of operations.*[88]

The influence of operationism on Money's formulation of gender explains his misstep between causality and prediction. Operationally, gender is simply the set of operations for its determination: the measurement and meaning of gender are one and the same. In the context of using sex assignment and rearing to predict an infant's gender, operationism suggests that gender is synonymous with the operations by which it is predicted, and that sex assignment and rearing actually *are* gender.

## The Promise of Feedback

Bridgman was no cybernetic theorist, and operationism was not cybernetics. Nevertheless, Money endeavored to synthesize operationism and cybernetics by using the concept of feedback. In his 1949 paper that referenced Bridgman and Wiener, Money outlined a model for the study of interrelated variables, which would structure his subsequent claims about gender. In the paper, Money addressed the question of how to investigate processes that were characterized by the absence of "cause and effect relationships" between "isolated variables."[89] The scientific answer, he contended, was to divide such processes into "operators amongst which the reciprocal relationships—that is, the feed-back relationships—can be determined."[90] Once a process was partitioned in this way, the goal was to identify the "intensity" with which each operator contributed to the feedback.[91] This was the blueprint for Money's prediction machine, anticipating his approach to the variables of sex. In fact, as an example of a "feed-back system," in 1949 Money presciently suggested "sexual role."[92] In later publications, Money elaborated on his understanding of both feedback and cybernetics. During the 1960s, he defined feedback as the process whereby "a signal produces a response that then itself becomes a signal to produce an alteration of the initial signal."[93] This definition, although inelegant, accurately represented cybernetic theory: in Wiener's words, feedback meant "controlling a system by reinserting into it the results of its past performance."[94] Money was also right, some twenty years later, to define cybernetics itself as regulation "by mutual feedback."[95] Yet, when Money made the unifying move of explaining gender's determination as an instance of feedback, he committed a serious mistake, as I will now argue.

Having delineated the prediction machine with which, Money claimed, "therapeutic plans" could be made for the "complete and predictable" differentiation of gender, he faced a challenge.[96] The challenge was to detail the exact relationship between gender and sex of assignment and rearing. The importance of specifying the nature of this relationship cannot be overstated: it was essential to the coherence of Money's treatment recommendations. Without clarity about the relationship between gender and sex of assignment and rearing, the recommendations would be mere speculation, unscientific by Money's own criteria. Feedback was a potential solution. Money had already used the concept to try to unite operationism and cybernetics in his early work, so he turned to feedback again when explaining the determination of gender in the prediction machine. Commenting on a report that mothers of female infants touched their children more often than mothers of male infants, and that female infants, in turn, tried more frequently than male infants to touch their mothers, Money distinguished two possible interpretations. By one interpretation, these behaviors were a "cause-effect riddle," soluble only by prioritizing either "the parents' gender-dimorphic expectancy, or the infants' gender-dimorphic activity." Characteristically, Money rejected that interpretation. The correct interpretation, in his view, was that expectancy and activity constituted "a feedback effect in gender-dimorphic behavior," or more concisely, a "gender-feedback effect." According to Money, this interpretation prioritized neither parent nor infant, but took instead their "interactional feedback" as the unit of analysis. Moreover, from the example of touching, Money extrapolated that all parent-child feedback was the product of evolution, and that its function was "the development of behavioral sexual dimorphism"—in other words, gender itself—in successive generations.[97] By Money's account, then, a feedback relationship existed between gender and sex of assignment and rearing.

A possible counterargument to Money's claim about feedback would be that the relationship between gender and sex of assignment and rearing is unidirectional. By reading determination as control, one could argue that no interaction exists, because gender is simply caused by sex of assignment and rearing. Indeed, Money once told his readers that the parental "concept of you as a boy or a girl, backed by everybody else in your world, pressed relentlessly upon you."[98] This would seem to describe not feedback, but straightforward domination of one variable by another. Whereas feedback, as Ashby put it, "exists between two parts when each affects the other," in the case of domination, "the action is one way only."[99] However, that is not my critique. Both feedback and domination are relationships between discrete variables. Neither relationship joined gender with sex of assignment and rearing in Money's prediction machine, because gender and sex of assignment and rearing were not really discrete. I think they had no relationship at all, for they were identical. Further, I submit that this explanation is the only way to take seriously the influence of operationism on Money's thought, recognizing the sameness of gender and the operations of its prediction.

The identity of the supposedly discrete variables is evident in the following passage from Money's 1972 book *Man and Woman, Boy and Girl*:

> Assignment of sex is not synonymous with the registration of sex on the birth certificate. Registration is a discrete act, whereas assignment becomes synonymous with rearing, insofar as a child is daily confronted with his boyhood, or her girlhood, in innumerable reaffirmations of assignment, including the gender forms of personal reference embedded in the nouns and pronouns of the language.[100]

In this passage, synonymy is established between assignment and rearing, which explains their treatment as a single variable throughout Money's work.[101] But synonymy is also established between rearing and gender, whereby the former is defined to include the latter—"the gender forms of personal reference" in "the nouns and pronouns of the language." Hence, in place of a relationship between assignment, rearing, and gender, there are merely "innumerable reaffirmations of assignment": neither feedback nor domination, but repetition, which Money elsewhere dubbed a "reiterative routine."[102] What is more, in another publication, Money amended the distinction made above between sex assignment and birth certificate registration, clarifying that

> By sex of assignment I meant quite literally the sex according to which a baby is classified and registered as either male or female as, for example, on the birth certificate, and according to which, on a day to day basis, the baby is referred to and reared as he or she.[103]

This comment unifies sex registration and assignment, before recapping the synonymy of assignment, rearing, and gender (the pronouns "he or she"), as well as underlining their repetitive "day to day" operation. Therefore, I think that Money was not describing a steady state in the homeostatic, dynamic sense, but rather a solid state, singular and annealed.

By arguing for the absolute synonymy of variables that appeared separate, I am taking a different course to other critics of Money's interactionism—including those discussed by Nikki Sullivan in chapter 1—who have proposed more complex, recursive accounts of the interactions that compose gender development.[104] My interest lies not in revealing gender to be more interactive and dynamic than Money thought, but in understanding its uniform solidity within his work. Sex of registration, sex of assignment, sex of rearing, and gender were solid because they were all synonymous with language, and specifically with sexually dichotomous language. The ring of synonyms ran as follows. In sex registration, Money noted, there was "no in-between position" on the birth certificate.[105] Correspondingly, in sex assignment, he wrote that there was "no pronoun for hermaphrodite."[106] In sex of rearing, in turn, Money stated that the label "boy" or "girl" had "tremendous force as a self-fulfilling prophecy," by directing "the full weight of society to one side or the other."[107] Finally, Money claimed, the "critical growing-up period for establishment of gender" was "also the one in which pronoun and proper name usage is established," such that once a child "has command of names, nouns and pronouns differentiating the sexes, a boy has a clear concept of himself as a boy, and a girl of herself as a girl."[108] The last remark closed the ring of synonyms, by defining gender as fluency in the same sexually dichotomous language of registration, assignment, and rearing. To Money, attaining such fluency meant becoming aware of one's genital appearance, and simultaneously comparing oneself to others "who use the same gender-differentiating nouns and pronouns."[109] It followed from this claim that surgery to disambiguate genitalia was a type of language lesson. Genital surgery facilitated the use of dichotomous nouns and pronouns in what Money called the "everyday" sense of "polar opposites, distinguished by the external sex organs."[110] So construed, any surgery that made an individual's external sex organs more readily distinguishable as female or male was beneficial to not only its recipient, but all users of language.

Normal gender development, then, meant becoming a child-sized sexologist, acquiring a language for sexual dimorphism just as Money had given to sexology the term "gender." Consequently, I contend that Money's treatment recommendations were tautological instead of cybernetic. In giving gender to sexology, Money had transplanted the term from "the vocabulary of philology and grammar."[111] Its philological and grammatical usage was (and remains) the differentiation, nominally by sex, of types of nouns and their associated parts of speech. Money's explanation for his terminological transplant was that gender allowed avoidance of a "dilemma of the pronouns" when addressing individuals who had intersex anatomies. By way of example, Money stated that it would have been wrong to address "a highly paid and nationally known female fashion model, with male chromosomal and gonadal sex" as "he or him, instead of she or her."[112] I agree with Money on that point, for reasons of moral recognition. Yet, if this scenario constituted a dilemma, it was dilemmatic by Money's own contrivance. Chromosomal and gonadal sex, like the other physical sex variables that Money demarcated, were by definition not gender. Conversely, in Money's view, an individual's "self-concept" was "by its very nature gender-differentiated."[113] So, it would have been nonsensical to distinguish a person's gender from the pronoun by which they were addressed.[114] I would argue further that the source of gender in philology and grammar reveals Money's prediction machine to be a tautological hoax: gender's independence from the physical variables of sex, and its synonymy with the sexually dichotomous language of registration, assignment, and rearing, was nothing other than the definition of gender. It is therefore very peculiar that on hundreds of occasions in his published work, Money drew parallels between gender development and the development of language, in statements such as "a gender role is established in much the same way as is a native language."[115] Appearing consistently without citations to literature on language learning, this parallel was neither a hypothesis nor a research finding: it was a comparison between a person's role in language and a person's role in language—a reiteration of the absolute solidity of sex registration, assignment, rearing, and gender.

## FROM CYBERNETICS TO PSYCHOANALYSIS

I have argued that Money's treatment recommendations were not cybernetic, despite his extensive engagement with cybernetic terminology and concepts. A question that follows from my critique is what Money's recommended treatments actually did, if they were neither predictive nor causal, but tautological. One interpretation, favored by biologist Milton Diamond among others, is that treatments simply supported or contradicted an individual's innate predisposition to identify as a given gender.[116] I will present a different interpretation, focusing on another hitherto unremarked account of communication and control in Money's work: the process of hypnotic suggestion. In this final section of the chapter, I will appraise Money's idiosyncratic explanation of gender development as "imprinting," and suggest that it was a cipher for hypnosis.

Together with binary sexual difference, one of the few dichotomies sanctioned by Money was the "opposition between the eradicable and the ineradicable."[117] He regarded this dichotomy as informative in identifying phenomena that become "permanently cemented into the brain," irrespective of their hereditary or environmental provenance.[118] To Money, one such phenomenon was gender, and another was language.[119] Those claims were of course tautological, but the mechanism of "cementing" differed nonetheless from the synonymy between sex registration, assignment, rearing, and gender. It concerned the enduring representation of such synonymy in an individual's brain—or, as Money sometimes put it more diffusely, in "the central nervous system."[120] In other words, it was a matter of how gender could be psychologized as a "conviction," felt by individuals to be permanent and involuntary. Understanding this mechanism was important for Money: if he were able to explain the persistence of gender in individuals without genital ambiguity, then he could theorize its clinical recreation. Although Money's training in psychology might have led him to reach for a psychoanalytic narrative of childhood identification with sexually dichotomous parent figures, he declined to use psychoanalytic theory on the grounds that its nomenclature would have hampered communication with physicians, and also that it could not explain the establishment of gender prior to the Oedipal stage, the start of which Money placed at age five.[121] Additionally, psychoanalysis would have entailed a theory of unconscious motivation which, as I discussed above, Money cast as outdated. These efforts to distance his work from psychoanalysis are notable because, like those critiqued by Lisa Downing in chapter 2, they coexisted with Money's use of psychoanalytic theory in sexology. So, as we will discover, Money's comments on hypnosis were distinctively psychoanalytic nevertheless.

## The Imprint of Gender

For a model of "the permanence, ineradicableness or irreversibleness" of development, Money turned to animal studies, in particular the work of Austrian zoologist Konrad Lorenz.[122] Researching the behavior of young birds, Lorenz had found that greylag goslings "unquestioningly accept the first living being whom they meet as their mother, and run confidently after him"—

even if the being in question were not a goose, but Lorenz. Mallard ducklings were more selective, requiring the zoologist to "quack like a mother mallard" while crouching below a certain height, before they would follow him as their mother.[123] Moreover, as Money noted in one of his commentaries on Lorenz's findings, "the ducklings responded to Lorenz as if he were their mother from that day onward," a long-term outcome that Money found "truly amazing."[124] Lorenz regarded the process to be so specific and enduring that in 1937 he named it "imprinting," after a German term for "stamping in."[125] Imprinting described the impressing of a stimulus (such as the mother's quack) onto an instinctive behavior (such as following). Put differently, it was the acquisition of an object for a reaction. The word "imprinting" was also intended to reflect the speed of the process, which as Money later glossed, "must take place during a critical or sensitive developmental period."[126] Some birds were susceptible to imprinting for just a few hours. Despite such brevity, the effects of imprinting always had an "absolute rigidity," according to Lorenz.[127] Imprinting, then, was another link in the genealogy of neologisms (flanked by homeostasis and operationism in the 1920s and cybernetics in the 1940s) to which Money added "gender" during the 1950s, and on which he drew.

In Lorenz's early work on imprinting, the zoologist cautioned that the process occurred only in birds; however, in the early 1950s, Lorenz suggested that some "irreversible" fixations in humans were similar to imprinting in their dependence on infantile experiences.[128] Money went further, asserting characteristically and on many occasions that both gender and language were "indelibly imprinted" in humans.[129] This formulation—an extrapolation from zoology that was Money's own—reconfigured two of the dichotomies that Money considered to be outdated. In imprinting, heredity and environment were amalgamated, with an environmental stimulus becoming integral to an innate behavior. Cause and effect were compressed into a critical time period, yet stretched over an animal's lifespan. Accordingly, Money hailed the critical period of development as an innovative twentieth-century temporal concept.[130] His uptake of imprinting in sexology therefore supported his vision of the discipline as an up-to-date successor to dichotomous thought. In fact, Money used imprinting to explain numerous phenomena studied by sexology, including "fetishism, exhibitionism, voyeurism, sadism, masochism, homosexuality and transvestism" as well as "falling in love."[131] Transposed into treatment recommendations for genitally ambiguous infants, imprinting's irreversibility could be employed "to great therapeutic advantage," Money claimed, in assuring the permanence of gender differentiation following surgical intervention.[132] By his account, such treatment replicated "the imprinting that takes place in anatomically normal people," generating the same psychological sense of one's gender (and sexuality) as "instinctive and inborn."[133] Despite the zoological origin of imprinting, it became for Money another way to talk about the plasticity of humanity—a capacity to receive impressions, common to all humans, irrespective of genital anatomy.

In claiming that gender and other phenomena were imprinted in humans, Money was using the term in a profoundly different way than Lorenz. Imprinting might suggest the imposition of a reaction onto an animal by an external stimulus, but that is not how it works. For example, it would be incorrect to think that a mallard duckling forever follows the creature that taught it to follow. Lorenz's point was that a duckling does not need to be taught to follow; it will do so instinctively. Imprinting is too fast and simple a process to include the learning of a behavior. What is imprinted, rather, is a stimulus for the activation of following behavior, whether exhibited at the moment of imprinting, or subsequently.[134] The stimulus is the behavior's "releaser," as Lorenz put it.[135] Understood this way around, instinctive behavior is actually inhibited most of the time, and performed only upon perception of its releaser. Lorenz explained such selectivity as the operation of an "innate perceptual structure," which acts as "a sort of filter letting through only sharply defined combinations of sensory data"—the quack and height of a mother mallard, for instance. Together, releaser and perceptual structure constitute what Lorenz named an "innate releasing mechanism."[136] He observed other innate releasing mechanisms too, such as birds preparing to take flight in reaction to the preparations of fellow flock members to do the same.[137] Imprinting is distinctive among innate releasing mechanisms because it refines the perceptual structure for the release of future behaviors. So, for a mallard duckling, the necessary stimulus for following becomes not just any quacking animal, but one in particular.

Money described all these aspects of avian imprinting in his writings, and did so mostly accurately. He noted the coexistence of "a genetic releasing mechanism" and "an environmental trigger or releaser," and the interdependence of their operation.[138] He explained that "a perceptual signal is matched to an innate releasing mechanism (IRM)," which "releases a behavioral pattern"—a statement that was largely right, but which implied the perceptual component to be exterior to the releasing mechanism.[139] Even so, Money stated correctly that in imprinting, the "perceptual pattern which originally activates an IRM gains exclusive power to operate that IRM."[140] However, something was conspicuously absent from Money's discourse on gender imprinting: the threshold. Familiar to Money from cybernetic theory, the concept of a threshold "for the release or inhibition of behavior" featured in many of his publications on gender and sexuality.[141] For instance, he categorized parental behavior as "sex shared but threshold dimorphic." By this, Money meant that women and men both exhibit parental behavior, but in response to the stimulus of a child, such behavior is evoked "more readily and more frequently in the mother than the father."[142] Money also informed his readers that "prenatal hormones lowered your threshold for some kinds of behavior," with the result that, for example, "it takes *less* stimulus to evoke your response as far as strenuous physical activity or challenging your peers is concerned."[143] He asserted similarly that a person's threshold for falling in love is lowered by the evocation of their "erotic response."[144] In view of Money's use of the threshold elsewhere, the term's unexplained absence from his account of gender imprinting is incongruous. It is revealing too, as I will now argue.

All innate releasing mechanisms entail the crossing of a threshold to release a behavior. This process defines them. Hence, in Lorenz's analysis of flocks taking flight, one bird's preparation to fly releases the same behavior in others nearby, in response to which more flock members prepare for flight, until "after an interval of mutual stimulation, the threshold of the whole reaction is so far lowered" that not only does the first bird take wing, but the flock follows.[145] In the case of imprinting with mallard ducklings, Lorenz's voice had to cross a threshold from human speech to quacking, traversing the ducklings' perceptual filter, in order to release their following behavior. If there were no threshold, there could be no imprinting, because instinctive behaviors would not require the perception of stimuli for their performance. I think that Money's omission of the threshold from his discourse on gender imprinting reveals the incommensurability between his use of imprinting and the zoological research on which he purported to draw. The omission of the threshold was not a failure by Money to describe fully the innate releasing mechanism of gender imprinting; it was a sign that gender imprinting was not an innate releasing mechanism in the first place. To be an innate releasing mechanism, gender would have to be released by a stimulus. However, if that were the case, then Money's axiom that humans are born "psychosexually neutral or undifferentiated" could not also be true, because there would be no gender to be released.[146] Since Money's treatment recommendations assumed that gender did not preexist its clinical and social determination, imprinting had to name something other than an innate releasing mechanism.

Gender was the object of imprinting. It was not released, but simply imposed. Although Money insisted that his concept of imprinting was identical to the zoological version, the two were unalike, because by Money's account, gender was something with which humans "become imprinted."[147] This was the equivalent of saying (incorrectly) that ducklings are imprinted with following behavior. All the same, Lorenz's term gave Money a vocabulary for the conviction and permanence that, in his view, characterized the experience of gender. It inspired him to use topographic metaphors derived from the connotations of "stamping in": Money described gender as "indelibly engraved," "imbedded and indelible," and "embedded firmly" once the critical period for imprinting had elapsed.[148] These metaphors, suggestive of plastic surgery on the psyche, gave the appearance of a psychological explanation for gender development. However, they were essentially restatements of the imprinting mechanism, construed by Money as an imposition onto the individual that left a permanent mark. On this point, my critique of Money's discourse on imprinting converges with my critique of his cybernetics. Money's version of imprinting was consistent with the solidity of sex registration, assignment, rearing, and gender. Both aspects of his work entailed a substitution of irrevocability and solidity for feedback and dynamics. Both demonstrated Money's failure to think cybernetically.

## Two Stories of Communication and Control

While Lorenz's research into imprinting predated Wiener's cybernetics, the zoologist had used the terminology of information transmission from the start: he likened the perceptual filter in an innate releasing mechanism to "a highly selective wireless receiving-station."[149] Lorenz later referenced cybernetics directly, remarking that it would be "very illuminating" to build a "cybernetic machine" for the replication of an innate releasing mechanism.[150] He did not elaborate, but I suggest that the reason for Lorenz's interest was that imprinting, by his account, was already cybernetic. It was a system of variables related by feedback. So, in his encounter with the ducklings, Lorenz's voice was a variable that switched from speaking to quacking, reciprocated by an alteration in another variable, the behavior of the ducklings. This was a feedback relationship rather than one-way domination, because Lorenz soon learned that he "could not cease from quacking for any considerable period without promptly eliciting the 'lost peeping' note in the ducklings," who thought themselves abandoned when they could no longer hear their mother.[151] The variables thereby acted upon each other. To paraphrase Wiener's definition of feedback, the results of the system's past performance were reinserted to control its future behavior: Lorenz had to keep imitating a mother mallard, and the system self-regulated into a new, steady state of quacking and following.

Money's recommendations for the imprinting of gender exemplified a theory of communication and control too, but not a cybernetic one. Because the goal of treatment, in Money's words, was for individuals with ambiguous genitalia to "grow up oblivious of ever having had a problem," the effects of gender imprinting needed to be not merely permanent, but also experientially "involuntary."[152] Had Money's version of imprinting been cybernetic, the experience of gender as involuntary might have been interpretable as ongoing homeostatic self-regulation. But since gender was imposed, the involuntary quality of gender could not be explained homeostatically. Instead, I argue, gender was unconsciously determined. In publications during the 1950s that have been assumed by previous commentators to be incidental to his work on gender, Money discussed the unconscious determination of behavior. He wrote that Sigmund Freud's early use of hypnosis "revealed the importance of events no longer within awareness" to symptoms of hysteria.[153] Moreover, Money described hypnosis as a way to plant, as well as to recover, unconscious information: because a hypnotized subject was "intensely suggestible," messages issued by the hypnotist to "learn and repeat a certain formula" were obeyed as yieldingly as those to recount a forgotten experience.[154] In hypnosis, therefore, a subject's behavior was determined unconsciously, and "in conformity with rules and regulations dictated by the hypnotist."[155] According to Money, such behavior could occur even after the conclusion of a hypnotic session.[156] In this regard, the session was a critical period that affected later behavior. However, unlike the critical period in zoology, hypnosis was a mechanism for the imposition of behavior onto an individual, because in the administration of hypnotic messages, communication and control were one.

In closing this chapter, I suggest that the therapeutic process that Money called gender imprinting was the establishment of gender by hypnosis. Money shared with Freud an interest in the relationship between behavior and events that were opaque to consciousness, because he intended for gender imprinting to be known to patients by its effects alone. Correspondingly, in Money's 1957 comments on the aftereffects of hypnosis, he distinguished between a hypnotist's message to behave in a certain way and the enactment of the message by a subject. Both message and enactment were "deferred," but the message remained unconscious to the subject even after its enactment.[157] This was more than a mere analogy for sex reassignment: to demonstrate a procedure for uncovering "the content of the deferred message," Money presented the case of an intersex patient. Named Louis, the patient had come under Money's study at age fourteen. By his mother's account, Louis had been reared as female until age four, during which time his name had been

Mary Louise. In interviews with Money, Louis did not speak about the reassignment or about the house in which he had lived between the ages of four and five. In response to questions concerning that year, Louis exhibited what Money called "signs to warn the investigator of a deferred message": he fell silent and still and hung his head. In an astonishingly disproportionate effort to uncover the deferred message of reassignment, Louis was given a dose of sodium amytal and interviewed again. The barbiturate sodium amytal, glossed by Money as a "hypnotic drug," had been used during World War II to treat shell shock.[158] Known colloquially in the 1920s and 1930s as a "truth serum," it was claimed by psychiatrists in the 1940s to cure the amnesia of overwhelming experiences that manifested in some soldiers as hysterical symptoms.[159] In the case of Money's teenage patient, the drug was administered with no therapeutic purpose, because Money already knew Louis's history from his mother. The sole aim of the procedure was to discover whether, as Money put it, Louis was "amnesiac for the message about his change of sex" or "simply too bashful to talk." After the sodium amytal, Louis did talk about his fifth year of life, but not about the reassignment from female to male. He described the house in which he had lived, but said that his name at the time had been Louis. Money interpreted this outcome as a failure to release a deferred message.[160] Yet, his patient's obdurate identification with the gendered noun "Louis" was also the right outcome for Money, because it signaled the success of gender imprinting. If reassignment were a hypnotic message that remained unconscious, then the experiential persistence of gender was its ongoing enactment—conscious and involuntary, all at once.

Or, perhaps, Louis just did not want to talk to John Money.

## NOTES

1. John Money, "Prenatal Hormones and Postnatal Socialization in Gender Identity Differentiation," in *Nebraska Symposium on Motivation*, ed. James K. Cole and Richard Dienstbier (Lincoln: University of Nebraska Press, 1973), 221–95, 223.

2. For example, John Money, Joan G. Hampson, and John L. Hampson, "Hermaphroditism: Recommendations Concerning Assignment of Sex, Change of Sex, and Psychologic Management," *Bulletin of the Johns Hopkins Hospital* 97 (1955): 284–300, 299; Money, "Prenatal Hormones," 223; John Money, "Unanimity in the Social Sciences with Reference to Epistemology, Ontology, and Scientific Method," *Psychiatry* 12 (1949): 211–21, 219.

3. Jennifer E. Germon, *Gender: A Genealogy of an Idea* (Basingstoke: Palgrave Macmillan, 2009), 32.

4. Germon, *Gender*, 46.

5. W. Ross Ashby, *An Introduction to Cybernetics* (New York: John Wiley, 1956), 53.

6. Money, "Unanimity in the Social Sciences," 213n2.

7. Table of contents, *Scientific American*, November 1948, 7.

8. Norbert Wiener, *Cybernetics; Or, Control and Communication in the Animal and the Machine*, 2nd ed. (Cambridge, MA: MIT Press, 1961), 11–12.

9. Norbert Wiener, "Cybernetics," *Scientific American*, November 1948, 14–19, 14.

10. Geof Bowker, "How to Be Universal: Some Cybernetic Strategies, 1943–70," *Social Studies of Science* 23 (1993): 107–27, 109.

11. Peter Galison, "The Ontology of the Enemy: Norbert Wiener and the Cybernetic Vision," *Critical Inquiry* 21 (1994): 228–66, 238, 240.

12. Wiener, *Cybernetics*, 11.

13. Walter B. Cannon, "Organization for Physiological Homeostasis," *Physiological Review* 9 (1929): 399–431, 400.

14. For example, W. Ross Ashby, *Design for a Brain: The Origins of Adaptive Behaviour*, 2nd ed. (London: Chapman and Hall, 1960), 100–121.

15. Ashby, *Design for a Brain*, 130–31.

16. Cannon, "Organization for Physiological Homeostasis," 400, 401.

17. Wiener, "Cybernetics," 14–15.

18. Wiener, "Cybernetics," 14.

19. Wiener, "Cybernetics," 15.

20. Franklin V. Taylor, review of *Cybernetics; Or, Control and Communication in the Animal and the Machine*, by Norbert Wiener, *Psychological Bulletin* 46 (1949): 236–37, 237.

21. John Money, "Psychologic Studies in Hypothyroidism," *Archives of Neurology and Psychiatry* 76 (1956): 296–309, 307; Wiener, "Cybernetics," 17; see also 18.

22. John Money, "Sexology: Behavioral, Cultural, Hormonal, Neurological, Genetic Etc.," *Journal of Sex Research* 9 (1973): 1–10, 5.

23. Money, "Sexology," 5; Wiener, "Cybernetics," 14.

24. John Money, "The Development of Sexology as a Discipline," *Journal of Sex Research* 12 (1976): 83–87, 86; Ashby, *Introduction to Cybernetics*, 4, 1.

25. John Money, *Sin, Science, and the Sex Police: Essays on Sexology and Sexosophy* (Amherst, NY: Prometheus, 1998), 15.

26. Walter B. Cannon, "The Body Physiologic and the Body Politic," *Science* 93 (1941): 1–10, 5.

27. Cannon, "Body Physiologic," 8.

28. Norbert Wiener, *The Human Use of Human Beings: Cybernetics and Society*, rev. ed. (Boston: Houghton Mifflin, 1954), 51–52.

29. Wiener, *Human Use of Human Beings*, 52; "Cybernetics," 18.

30. Money, "Unanimity in the Social Sciences," 216; Wiener, *Cybernetics*, 11.

31. John L. Hampson, Joan G. Hampson, and Money, "The Syndrome of Gonadal Agenesis (Ovarian Agenesis) and Male Chromosomal Pattern in Girls and Women: Psychologic Studies," *Bulletin of the Johns Hopkins Hospital* 97 (1955): 207–26, 225.

32. John Money, *Gendermaps: Social Constructionism, Feminism, and Sexosophical History* (New York: Continuum, 1995), 136.

33. Ashby, *Introduction to Cybernetics*, 1.

34. Money, "Prenatal Hormones," 223.

35. Alan Sheldon, "Cybernetics and Medical Care: Some Recent Literature," *Social Science and Medicine* 3 (1969): 281–88, 283.

36. John Money, *The Psychologic Study of Man* (Springfield, IL: Charles C. Thomas, 1957), 48.

37. John Money, "Sexual Dimorphism and Homosexual Gender Identity," *Psychological Bulletin* 74 (1970): 425–40, 425; John Money and Anke A. Ehrhardt, *Man and Woman, Boy and Girl: The Differentiation and Dimorphism of Gender Identity from Conception to Maturity* (Baltimore: Johns Hopkins University Press, 1972), xi.

38. Money, "Prenatal Hormones," 223; Wiener, "Cybernetics," 14.

39. Ashby, *Introduction to Cybernetics*, 191.

40. Money, "Prenatal Hormones," 223.

41. Money, "Development of Sexology," 86.

42. Money, "Prenatal Hormones," 223.

43. John Money and Mark Schwartz, "Biosocial Determinants of Gender Identity Differentiation and Development," in *Biological Determinants of Sexual Behaviour*, ed. John B. Hutchison (Chichester: John Wiley, 1978), 765–84, 782; Arturo Rosenblueth, Norbert Wiener, and Julian Bigelow, "Behavior, Purpose and Teleology," *Philosophy of Science* 10 (1943): 18–24, 24.

44. Money, *Psychologic Study of Man*, 33.

45. Money, "Sexology," 10; see also Money, "Prenatal Hormones," 222–23.

46. Ashby, *Introduction to Cybernetics*, 66.

47. Cannon, "Organization for Physiological Homeostasis," 422.

48. Money, *Psychologic Study of Man*, 3.

49. Money, *Psychologic Study of Man*, 5.

50. Ashby, *Introduction to Cybernetics*, 1; Money, *Sin, Science, and the Sex Police*, 283; Money, *Psychologic Study of Man*, v.

51. Money, "Sexology," 4.

52. Anatol Rapoport, foreword to *Modern Systems Research for the Behavioral Scientist: A Sourcebook*, ed. Walter Buckley (Chicago: Aldine, 1968), xiii–xxii, xix; Orit Halpern, "Dreams for Our Perceptual Present: Temporality and Interactivity in Cybernetics," *Configurations* 13 (2005): 283–319, 291.

53. Wiener, *Human Use of Human Beings*, 61.

54. David A. Rubin, "'That Unnamed Blank That Craved a Name': A Genealogy of Intersex as Gender," *Signs* 37 (2012): 883–908, 900; see also 892, 897.

55. Money and Ehrhardt, *Man and Woman*, 153.

56. John Money, "Hermaphroditism, Gender, and Precocity in Hyperadrenocorticism: Psychologic Findings," *Bulletin of the Johns Hopkins Hospital* 96 (1955): 253–64, 258; John Money, preface to *Sex Research: New Developments*, ed. John Money (New York: Holt, Rinehart and Winston, 1965), ix.

57. Money and Ehrhardt, *Man and Woman*, 153, 179. The metaphor of transmission can be read both cybernetically and epidemiologically. Money used it predominantly in the first sense, although at least once he claimed that parental equivocation over a child's gender could

be transmitted "as contagiously as though it were rubella" (Money and Ehrhardt, *Man and Woman*, 15). The cybernetic and epidemiological connotations do not annul each other, but signal the coexistence of discourses in Money's work.

58. Germon, *Gender*, 22; Rubin, "That Unnamed Blank," 887.

59. John Money, "Gender: History, Theory and Usage of the Term in Sexology and Its Relationship to Nature/Nurture," *Journal of Sex and Marital Therapy* 11 (1985): 71–79, 73.

60. Money, "Sexology," 10.

61. John Money, Joan G. Hampson, and John L. Hampson, "An Examination of Some Basic Sexual Concepts: The Evidence of Human Hermaphroditism," *Bulletin of the Johns Hopkins Hospital* 97 (1955): 301–19, 301. Even though individual elements could sometimes be composed of female and male characteristics together (such as gonads containing ovarian and testicular tissue), the partitioning of sex allowed Money to categorize the majority of intersex conditions as unambiguously dichotomous by most criteria—whereby "at least one" element was "contradictory of the remainder," insofar as dichotomously female and male elements coexisted (John Money, "Hermaphroditism," in *The Encyclopaedia of Sexual Behaviour*, vol. 1, ed. Albert Ellis and Albert Abarbanel [London: Heinemann, 1961], 472–84, 473; hereafter "Hermaphroditism [encyclopedia entry]").

62. John Money, *A First Person History of Pediatric Psychoendocrinology* (New York: Kluwer Academic/Plenum, 2002), 35.

63. Money, Hampson, and Hampson, "Hermaphroditism: Recommendations Concerning Assignment of Sex," 299.

64. Money, Hampson, and Hampson, "An Examination of Some Basic Sexual Concepts," 301; John Money, Joan G. Hampson, and John L. Hampson, "Imprinting and the Establishment of Gender Role," *American Medical Association Archives of Neurology and Psychiatry* 77 (1957): 333–36, 333.

65. Money, "Prenatal Hormones," 292; see also note 126 in chapter 3, above.

66. Hampson, Hampson, and Money, "Syndrome of Gonadal Agenesis," 225.

67. Money, Hampson, and Hampson, "Hermaphroditism: Recommendations Concerning Assignment of Sex," 299; Money, Hampson, and Hampson, "An Examination of Some Basic Sexual Concepts," 301.

68. John Money, glossary in *Sex Research: New Developments*, ed. John Money (New York: Holt, Rinehart and Winston, 1965), 235–50, 248; John Money, "Prenatal Hormones," 292.

69. Money, Hampson, and Hampson, "Imprinting," 333.

70. Money, Hampson, and Hampson, "An Examination of Some Basic Sexual Concepts," 301.

71. Money and Schwartz, "Biosocial Determinants," 781.

72. Ashby, *Introduction to Cybernetics*, 39, 40.

73. Ashby, *Introduction to Cybernetics*, 40, 16.

74. Money, *Sin, Science, and the Sex Police*, 59; Ashby, *Design for a Brain*, 14 (Ashby's italics).

75. Ashby, *Introduction to Cybernetics*, 126.

76. Hampson, Hampson, and Money, "Syndrome of Gonadal Agenesis," 225; John Money, "Components of Eroticism in Man: Cognitional Rehearsals," in *Recent Advances in Biological*

*Psychiatry: Including a Havelock Ellis Centenary Symposium on Sexual Behavior*, ed. Joseph Wortis (New York: Grune and Stratton, 1960), 210–25, 216.

77. John Money, "Propaedeutics of Diecious G-I/R: Theoretical Foundations for Understanding Dimorphic Gender Identity/Role," in *Masculinity/Femininity: Basic Perspectives*, ed. June Machover Reinisch, Leonard A. Rosenblum, and Stephanie A. Sanders (New York: Oxford University Press, 1987), 13–28, 16.

78. Money, "Hermaphroditism [encyclopedia entry]," 479–80.

79. Money, Hampson, and Hampson, "An Examination of Some Basic Sexual Concepts," 319. In cases where an individual's gender developed ambiguously, Money still traced the outcome to this variable, asserting that "ambiguity of rearing, especially if reinforced by ambiguity of body appearance, leads to an ambiguous gender identity" (John Money, "Critique of Dr. Zuger's Manuscript," *Psychosomatic Medicine* 32 [1970]: 463–65, 464).

80. Money and Schwartz, "Biosocial Determinants."

81. Money, "Hermaphroditism," 264, 258. Once, Money claimed that "the sex of assignment and rearing does not automatically and mechanistically determine the gender role and orientation" (Money, Hampson, and Hampson, "Imprinting," 334–35). This remark appeared contrary to his treatment recommendations. However, to say that gender is not determined "automatically and mechanistically" by sex of assignment and rearing leaves open the possibility that gender could be determined by sex of assignment and rearing in some other fashion—such as via a feedback relationship.

82. Money, "Hermaphroditism," 257; Money, Hampson, and Hampson, "Imprinting," 333.

83. Money, Hampson, and Hampson, "Imprinting," 333.

84. As Money's treatment recommendations were disseminated, the ambivalence in his writings enabled a shift of accent from prediction to control. For example, gynecologist Edwin C. Hamblen attributed to Money and his colleagues the finding that "the sex of assignment and rearing determines to a great degree the gender and orientation of the individual." Hamblen hereby accented the causal, controlling aspect of determination, without its ambivalent prognosticative connotations ("The Assignment of Sex to an Individual: Some Enigmas and Some Practical Clinical Criteria," *American Journal of Obstetrics and Gynecology* 74 [1957]: 1228–44, 1230).

85. John Money, *Venuses Penuses: Sexology, Sexosophy, and Exigency Theory* (Amherst, NY: Prometheus, 1986), 7.

86. Money, *Venuses Penuses*, 7.

87. Money, "Unanimity in the Social Sciences," 212.

88. Percy Williams Bridgman, *The Logic of Modern Physics* (New York: Macmillan, 1927), 5 (Bridgman's italics).

89. Money, "Unanimity in the Social Sciences," 213.

90. Money, "Unanimity in the Social Sciences," 214.

91. Money, "Unanimity in the Social Sciences," 213.

92. Money, "Unanimity in the Social Sciences," 219.

93. Money, glossary in *Sex Research*, 239.

94. Wiener, *Human Use of Human Beings*, 61.

95. John Money, *Gay, Straight, and In-Between: The Sexology of Erotic Orientation* (New York: Oxford University Press, 1988), 116.

96. Money and Ehrhardt, *Man and Woman*, 234.

97. Money and Ehrhardt, *Man and Woman*, 180. Elsewhere, Money described "feedback from the community" to a sex-reassigned child, following explanations of the reassignment given to the community by the child's family (John Money, Reynolds Potter, and Clarice S. Stoll, "Sex Reannouncement in Hereditary Sex Deformity: Psychology and Sociology of Habilitation," *Social Science and Medicine* 3 [1969]: 207–16, 212).

98. John Money and Patricia Tucker, *Sexual Signatures: On Being a Man or a Woman* (Boston: Little, Brown, 1975), 87.

99. Ashby, *Introduction to Cybernetics*, 53, 54.

100. Money and Ehrhardt, *Man and Woman*, 15.

101. After their collaboration with Money had ended, the psychologists Joan and John Hampson stated that "the term *gender role* is not identical and synonymous with the term *sex of assignment and rearing*" (John L. Hampson and Joan G. Hampson, "The Ontogenesis of Sexual Behavior in Man," in *Sex and Internal Secretions*, 3rd ed., vol. 2, ed. William C. Young [Baltimore: Williams and Wilkins, 1961], 1401–32, 1406n5). Money did not make the same claim.

102. John Money, *Sex Errors of the Body: Dilemmas, Education, Counseling* (Baltimore: Johns Hopkins Press, 1968), 61.

103. Money, *A First Person History*, 35.

104. For example, Ruth G. Doell and Helen E. Longino, "Sex Hormones and Human Behavior: A Critique of the Linear Model," *Journal of Homosexuality* 15 no. 3/4 (1988): 55–78, 74; Anne Fausto-Sterling, *Sexing the Body: Gender Politics and the Construction of Sexuality* (New York: Basic, 2000), 242–43; Rebecca M. Jordan-Young, *Brain Storm: The Flaws in the Science of Sex Differences* (Cambridge, MA: Harvard University Press, 2010), 286; Lesley Rogers, "The Ideology of Medicine," in Dialectics of Biology Group, *Against Biological Determinism*, ed. Steven Rose (London: Allison and Busby, 1982), 79–93, 84.

105. Money, *Sin, Science, and the Sex Police*, 346.

106. Money, *A First Person History*, 35. Money proposed that the non-gender specific pronoun "o" should be adopted from Turkish into English, but as a replacement for third-person formulations such as "he or she" (Money and Tucker, *Sexual Signatures*, 117–18).

107. Money and Tucker, *Sexual Signatures*, 86.

108. Money, *Psychologic Study of Man*, 51; John Money, "Determinants of Human Gender Identity/Role," in *Handbook of Sexology*, ed. John Money and Herman Musaph (Amsterdam: Elsevier/North Holland Biomedical, 1977), 57–79, 65.

109. Money and Schwartz, "Biosocial Determinants," 781.

110. Money, *Sin, Science, and the Sex Police*, 346. Genital appearance was not an extralinguistic reference point for the ring of synonyms, because Money described genitalia as a kind of language—"the sign, above all others" of gender (Money, "Hermaphroditism," 257).

111. Money, *A First Person History*, 35.

112. Money, *A First Person History*, 35.

113. Money and Ehrhardt, *Man and Woman*, 176.

114. For transsexual individuals, coming out does often involve distinguishing one's gender from the pronoun by which one has hitherto been addressed, with a view to recognition by the correct pronoun from thereon. But Money's "dilemma" was an instance of intersex, not transsexuality; and as Milton Diamond has pointed out, Money's gender theory struggled to account for the discrepancy between gender and sex of assignment and rearing in transsexual individuals (Milton Diamond, "Pediatric Management of Ambiguous and Traumatized Genitalia," *Journal of Urology* 162 [1999]: 1021–28, 1022).

115. Money, Hampson, and Hampson, "Hermaphroditism: Recommendations Concerning Assignment of Sex," 285. Money's first published articulation of this parallel was in "Hermaphroditism," 258.

116. Milton Diamond, "A Critical Evaluation of the Ontogeny of Human Sexual Behavior," *Quarterly Review of Biology* 40 (1965): 147–75, 151; see also, for example, Christopher J. Dewhurst and Ronald R. Gordon, *The Intersexual Disorders* (London: Baillière, Tindal and Cassell, 1969), 148.

117. Money, *Psychologic Study of Man*, 48.

118. Money, *Sin, Science and the Sex Police*, 55.

119. Money, *Psychologic Study of Man*, 51.

120. Money and Schwartz, "Biosocial Determinants," 766.

121. Money, *Venuses Penuses*, 8; Money and Ehrhardt, *Man and Woman*, 184. Money's dating of the Oedipal stage demonstrated a selective reading of psychoanalytic theory; in 1945, Melanie Klein had suggested that the Oedipus complex begins "during the first year of life," which would have accorded with Money's chronology of gender development ("The Oedipus Complex in the Light of Early Anxieties" [1945], in *Contributions to Psycho-Analysis 1921–1945* [London: Hogarth, 1968], 339–90, 378).

122. Money, *Psychologic Study of Man*, 52.

123. Konrad Z. Lorenz, *King Solomon's Ring: New Light on Animal Ways* [1949], trans. Marjorie Kerr Wilson (London: Routledge, 2002), 40–41.

124. Money, Hampson, and Hampson, "Imprinting," 335. Other substantial accounts of Lorenz's work are Money, *Psychologic Study of Man*, 52–53, and Money, "Components of Eroticism," 215–16.

125. Konrad Z. Lorenz, "The Companion in the Bird's World," *Auk* 54 (1937): 245–73, 262.

126. Money and Ehrhardt, *Man and Woman*, 286.

127. Lorenz, "Companion in the Bird's World," 264.

128. Lorenz, "Companion in the Bird's World," 266; Lorenz in *Discussions on Child Development: A Consideration of the Biological, Psychological, and Cultural Approaches to the Understanding of Human Development and Behaviour*, vol. 1, ed. J. M. Tanner and Bärbel Inhelder (London: Tavistock, 1956), 117.

129. Money, Hampson, and Hampson, "An Examination of Some Basic Sexual Concepts," 309–10. Another example of this assertion appears in John Money, "Psychosexual Differentiation," in *Sex Research: New Developments*, ed. John Money (New York: Holt, Rinehart and Winston, 1965), 3–23, 12.

130. John Money, "Psychologic Approach to Psychosexual Misidentity with Elective Mut-

ism: Sex Reassignment in Two Cases of Hyperadrenocortical Hermaphroditism," *Clinical Pediatrics* 7 (1968): 331–39, 339.

131. Money, "Components of Eroticism," 218, 219.

132. Money and Ehrhardt, *Man and Woman*, 178.

133. Money, "Components of Eroticism," 218.

134. Lorenz, "Companion in the Bird's World," 265.

135. Lorenz, "Companion in the Bird's World," 249.

136. Lorenz in Tanner and Inhelder, *Discussions*, 110–11.

137. Lorenz, "Companion in the Bird's World," 257.

138. Money, "Psychologic Studies in Hypothyroidism," 339.

139. Money, "Components of Eroticism," 224.

140. Money, "Components of Eroticism," 216.

141. Money and Schwartz, "Biosocial Determinants," 775.

142. Money, "Propaedeutics of Diecious G-I/R," 20.

143. Money and Tucker, *Sexual Signatures*, 90, 78.

144. John Money and John G. Brennan, "Heterosexual vs. Homosexual Attitudes: Male Partners' Perception of the Feminine Image of Male Transexuals," *Journal of Sex Research* 6 (1970): 139–209, 207.

145. Lorenz, "Companion in the Bird's World," 257.

146. John Money, "Cytogenetic and Psychosexual Incongruities with a Note on Space-Form Blindness," *American Journal of Psychiatry* 119 (1963): 820–27, 820.

147. Money and Ehrhardt, *Man and Woman*, 178; Money, "Hermaphroditism [encyclopedia entry]," 477.

148. Money, Hampson, and Hampson, "An Examination of Some Basic Sexual Concepts," 310; Money, *Psychologic Study of Man*, 51; Money and Schwartz, "Biosocial Determinants," 779.

149. Lorenz, "Companion in the Bird's World," 247.

150. Lorenz in Tanner and Inhelder, *Discussions*, 112.

151. Lorenz, "Companion in the Bird's World," 268. Later, the ducklings regarded Lorenz as their mother while he was silent; however, that was the result of associative learning, not imprinting ("Companion in the Bird's World," 268, 264). In a related critique of Money's account of imprinting, Diamond has suggested that gender role is learned rather than imprinted ("A Critical Evaluation," 165–66).

152. Money, *Sex Errors*, 43; Money, "Components of Eroticism," 218.

153. John Money, "Observations Concerning the Clinical Method of Research, Ego Theory, and Psychopathology," *Psychiatry* 14 (1951): 55–56, 57.

154. Money, *Psychologic Study of Man*, 38.

155. Money, *Psychologic Study of Man*, 39.

156. Money, *Psychologic Study of Man*, 39.

157. Money, *Psychologic Study of Man*, 39–40.

158. Money, *Psychologic Study of Man*, 40.

159. Alison Winter, "The Making of 'Truth Serum,' 1920–1940," *Bulletin of the History of Medicine* 79 (2005): 500–533, 502; William Sargant and Eliot Slater, "Acute War Neurosis,"

*Lancet* 236 (1940): 1–2, 2. Hypnotic and traumatic states both involve what the cultural critic Ruth Leys has called an "openness to impressions or identifications occurring prior to all self-representation and hence to all rememoration" (*Trauma: A Genealogy* [Chicago: University of Chicago Press, 2000], 32). Arguably then, if gender imprinting is hypnotic, it is also intentionally traumatic in its suddenness and involuntary persistence. I explore this possibility in "Intersex Treatment and the Promise of Trauma," in *Gender and the Science of Difference: Cultural Politics of Contemporary Science and Medicine*, ed. Jill A. Fisher (New Brunswick, NJ: Rutgers University Press, 2011), 147–63.

160. Money, *Psychologic Study of Man*, 40.

CHAPTER 5

# Reorienting Transsexualism: From Brain Organization Theory to Phenomenology

*Nikki Sullivan*

There is little doubt that John Money is a key figure in the history of transsexuality. Along with his colleagues Howard Jones Jr. and Milton Edgerton,[1] Money played a pivotal role in the establishment of America's first Gender Identity Clinic at Johns Hopkins University in 1966. As Joanne Meyerowitz notes, the clinic "accorded a certain professional legitimacy to sex reassignment,"[2] becoming the site of the first complete genital surgery to be carried out on an MTF in the United States.[3] Within months of the news becoming public, the clinic's staff "had performed surgeries on ten transsexual patients, five FTMs and five MTFs,"[4] and had received requests from over a hundred other individuals.[5] From the outset, Money was at the forefront of the push for public acceptance of reassignment procedures, arguing that "conversion surgery" is medically and socially justifiable in as much as it constitutes "an investigative, ameliatory therapy" for what is "a developmental anomaly,"[6] rather than, as some claimed, a sinful and/or immoral lifestyle choice. But despite his considerable contribution to the research and treatment of transsexualism, Money's work is rarely discussed in any detail in the writings of contemporary trans theorists and/or activists.[7]

There are a number of explanations one might offer for Money's fraught position vis-à-vis trans theory and politics, the most obvious being the construction of Money by John Colapinto and others as an exemplar of Gloucester's gods who "kill us for their sport" in *King Lear*, or Mary Shelley's Dr. Frankenstein whose "game of science fiction"[8] is ruthlessly self-serving and medically unethical. Indeed, the desire to distance oneself from Money's work in the midst of increasing criticism of the role it has played in what is now largely regarded (in some quarters at least) as the surgical mutilation of intersex children, is hardly surprising. Such a move is, in my opinion, problematic insofar as it obscures the fact that Money's model of psychosexual development (and the establishment of G-I/R) that underpins the medical model of intersex, has, at least historically, been key to the justification and popularization of reassignment procedures associated with transsexualism (as a medical disorder). Rather than dismissing or championing the contribution Money made to "transsexualism," the aim of this chapter is to provide a critical analysis of the tensions at work in Money's research and writings in this area.

## GENDERMAPPING

As I explained in chapter 1, Money coined the term "gendermap" to refer to "the entity, template or schema within the mind and brain (mindbrain) unity that codes masculinity, femininity, and androgyny."[9] The development of the gendermap occurs, according to Money, in an orderly sequential manner: there is, phylogenetically, "one road with a number of forks where each of us turns in either the male or the female direction.[10] You become male or female by stages,"[11] and through a developing sense of identification and complementation. In short, G-I/R is not wholly biologically determined at or prior to birth, but rather, evolves both pre- and postnatally. However, as I noted in chapter 1, for Money G-I/R is not therefore radically plastic: while flexible in early life, G-I/R becomes increasingly resolute as the individual matures, hence the claim that sex assignment procedures should be carried out as early as possible in the case of intersex children. Money writes, "Formations that are socially constructed become welded into the total structure [of the gendermap and its polarized schema: feminine/masculine; me/thee] where they may become as fixed and immutable as if they had been, to use common parlance, exclusively genetic or biological and nonsocial in origin."[12] Given this, Money and Tucker argue—again deploying the analogy of the road—that behaving in a way that is not in keeping with one's schema feels uncomfortable, unnatural, even abhorrent. "Looking at the schemas in this way may help . . . to understand the problems of transsexuals. As long as society makes them follow the traffic pattern for their anatomy and label, they are under the strain of constantly running lights that their gender identity insists are red."[13] And as Money repeatedly notes, no amount of psychotherapy seems able to change this.

So what exactly is it that produces the disjunction between body and self that Money claims preoperative transsexuals experience? As Money sees it, transsexualism is the end product of the process of psychosexual differentiation gone awry. In order to avoid "any dogma of causality,"[14] Money suggests that "a weakness or maldevelopment at any one stage of the progression may not directly induce a specific defect in the final identity. It does, however, introduce an instability in the structure, making each subsequent developmental level more vulnerable to defect."[15] What interests Money, then, is less the exact etiology of transsexualism and more the fact that it constitutes a form of "gender transposition" or "gender cross-coding" in which the schemas (feminine/masculine, identification/complementation, me/thee) change place such that one's gendermap is discordant with the natal sex of the external genitalia. In conceiving it in this way, Money constitutes transsexualism as ontologically equivalent (although different in expression) to homosexuality and transvestism.[16] This is perhaps best explained by turning to two companion pieces that Money coauthored in 1969: "Sexual Dimorphism and Dissociation in the Psychology of Male Transsexuals" (written with Clay Primrose), and "Sexual Dimorphism in the Psychology of Female Transsexuals" (written with John G. Brennan). But before we do this it is worth remembering that, for Money, sex is unquestionably naturally dimorphic, and sex and sexuality are inextricable from, or, more precisely, are aspects of, G-I/R. In other words, the developmental process of gendermapping that affects an individual's sense of self as male, female, both, or neither, simultaneously codes one's "lovemap as masculine, feminine, or bisexual."[17]

## SLUGS AND SNAILS, SUGAR AND SPICE, TOMBOYS AND SISSIES

Unlike intersexuality, which is largely believed to be visibly self-evident,[18] transsexualism, argues Money's colleague (and ex-student) Richard Green, is a behavioral diagnosis that "does not carry with it the precision frequently found in the medical and surgical sciences. There are no x-rays, tissue biopsies, or bacterial cultures available for the analysis of cross-gender behavior."[19] Consequently, in order for the clinician to maintain the (exclusory) position of the subject of (clinical) knowledge, he must somehow be able to determine objectively the "truth" of the patient's psychosexual status, and, in turn, their suitability for surgery. As Sandy Stone puts it, "professionally speaking, a test or a differential diagnosis was needed for transsexualism that did not depend on anything as simple and subjective as feeling that one was in the wrong body. The test needed to be objective, clinically appropriate, and repeatable."[20] Money believed he had found such a test in the Guilford-

Zimmerman Temperament Survey's masculinity-femininity scale. The test is comprised of true-false statements about behavior patterns that Money claims "are traditionally sex-differentiated":[21] for example, "you can look at snakes without shuddering"—to which those with "masculine [mind]brain organization" will (allegedly) answer, "true," and those with "feminine [mind]brain organization" will no doubt shudder and tick "false" before quickly moving on to more palatable questions about music, literature, dress designing, and of course, crying.[22]

In the two papers on sexual dimorphism and dissociation Money and his coauthors report that the questions used in the interviews with MTFs and FTMs, respectively, focus on five key signifiers of G-I/R: namely, the physical expression of aggression in childhood, genitopelvic functioning, imagery during masturbation and/or sexual intercourse, postoperative experience of phantom penis (in the case of MTFs) or phantom breasts and womb (in FTMs), and degrees of maternalism (in MTFs) or paternalism (in FTMs).[23] In order to be deemed suitable candidates or successful ambassadors for sex reassignment surgery (SRS), the individuals interviewed were required to demonstrate a coherent and firmly established G-I/R, and a clear rejection of the "opposite-sex model [that] exist[s] in the brain as a neurocognitional entity."[24]

In regard to the first of the behavioral issues mentioned above, Money and Primrose report that the fourteen MTF interviewees (whom they tellingly refer to as "male transsexual patients"[25]) "invariably . . . recalled an aversion to fighting, to 'boys' competitive games, and to rough, outdoor activities. . . . They much preferred the security of the home and little girl activities."[26] As a result, all of the interviewees had been "labeled 'sissy' by their peers."[27] The findings of the study of the six FTM transsexuals on file in the Psychohormonal Research Unit at Johns Hopkins Hospital in 1969 are similar; Money and Brennan write:

> Without exception, all of the patients reported that, as girls, they were considered to be extremely active. Four often engaged in fighting, usually with boys; one reported wrestling with and often beating boys. . . . All were considered to be tomboys and played boys' games and five had boys' ambitions. Four dislike and claimed never to have played with dolls; two played with them, but were never really interested in doing so.[28]

These behavioral traits are interpreted by Money and his coauthors as (unmediated, empirical) evidence of feminine brain organization and masculine brain organization, respectively,[29] thus supporting the hypothesis they set out to prove: namely, that transsexualism is an effect of "gender transposition" or "gender cross-coding." One might equally well argue, however, that gender cross-coding is a truth effect of a particular way of seeing, of a situated optics, a claim I will return to in the final section of the chapter.

Similarly, Money and Primrose's analysis of the "genitopelvic functioning" of those they interview begins with the heteronormative premise that "love play and other behavior attendant upon and including the copulatory act, is sexually dimorphic. . . . The origin of this dimorphism lies in the physical fact that, for fertilization alone, the male must have and hold an erection, while the female need only be receptive."[30] As I said in chapter 1, Money argues that irreducible G-I/R differences (the male's capacity to impregnate and the female's to menstruate, gestate, and lactate) subtend gender stereotypes, which then provide a continued (and necessary) framework for complementation, and thus for the continuation of the species. Unsurprisingly, the "genitopelvic functioning" of Money and Primrose's interviewees is interpreted as supporting this thesis. Money and Primrose write, the MTFs'

> conception of sexually dichotomous behavior is defined in conformity with . . . [gender] stereotypes, not their violations. . . . Eight of the eleven patients . . . who had had sexual relations . . . [had] participated in anal intercourse . . . [and/or] interfemoral intercourse as the insertee. Except for the three patients who had an episode of being married as a male, there was no incidence of coital insertion of the transsexuals' own penis into any orifice of the partner's anatomy.[31]

This is in keeping with Money's view of sexual mounting as typically male behavior, and "lordosis" (sexual presenting) as predominantly female behavior—a point I will return to in due course.

According to Money and Primrose, their MTF interviewees proved to be more aroused by touch than by visual stimulation and to frame their arousal in romantic terms. They also allegedly claimed to experience no penile pleasure, and, in some cases, to find their preoperative genitalia abhorrent. While Money and Primrose took the interviewees' responses to the questions posed at face value—not necessarily for humanitarian or political reasons, but more particularly, because such accounts of gender stereotypical behavior supported their hypothesis of gendermap cross-coding—Stone has since argued that in order to gain access to the medical procedures they required, trans people who attended gender identity clinics in this period quickly learned to tell the "right" stories, to perform appropriately in the "gender of choice."[32] She argues, for example, that

> physical men who lived as women and who identified themselves as transsexuals, as opposed to male transvestites for whom erotic penile sensation was permissible, could not experience penile pleasure. In the 1980s there was not a single preoperative male-to-female transsexual from whom data was available who experienced genital sexual pleasure while living in the "gender of choice." . . . To acknowledge so natural a desire [as the desire to masturbate] would be to risk "crash landing": that is, "role inappropriateness" leading to disqualification [from the program].[33]

Lacking the acute critical perspective Stone provides, Money and Primrose nevertheless seem to demonstrate *some* awareness that the "feminine sexual function" of which the MTF interviewees speak may not simply be the effect or expression of a "purely" biological cause when they ask "which claims precedence, the created image of feminine sexual function, or the change in neuropsychological functioning?"[34] Given Money's claim that his theory of G-I/R is interactionist one might assume that he would consider this chicken-and-egg question to be both unanswerable and misguided. And yet, ultimately, this proves not to be the case, as I will demonstrate in the next section, in which I discuss brain organization theory.

Money's characterization of transsexuals, transvestites, and homosexuals as "gender-transposed" is, as the companion studies that I've discussed show, based on an interpretation of the (gendered) behavior of such individuals as incongruent (in varying ways and degrees) with their physiology. This is particularly so in regard to erotic behavior: for example, Money sees "male bodied" individuals who are consistently more aroused by being penetrated than they are by penetrating, as "feminine," whereas sexual behavior that is initiatory, genitally focused, visually driven, and largely independent of concerns for love and intimacy is regarded as "masculine." Such interpretations (and the classifications they enable) are clearly far from objective, as I argued in chapter 1.[35] Moreover, they seem—to me, although perhaps not to Money—to contradict Money's oft-made claim that stereotyped gender roles are contextually specific and changing. But rather than strategically foregrounding tensions such as these in the hope that they might prove fruitful, Money's turn to "sex hormones" in order to explain "sexual brain differentiation,"[36] and the increasing resoluteness (or "nativization") of

G-I/R, undermines what I see as the radical potential of his work on gender acquisition.

As I mentioned earlier, the developmental model of G-I/R as increasingly nativized is foundational to Money's prescribing of SRS as an appropriate treatment for transsexualism (as a developmental anomaly). However, given that cross-gender behavior cannot be easily proven to have an indexical relation to some sort of visible pathogenic structure, the case for SRS was a more difficult one for Money and his colleagues at Hopkins to make than was the case for sex assignment in intersex children.[37] Consequently, Money drew on two existing fields of research in order to substantiate his position vis-à-vis transsexualism and SRS: namely, his own study of intersexuality and the research that brain organization theorists had carried out on animals.

## BRAIN ORGANIZATION THEORY: ANIMAL CRACKERS AND "NATURE'S OWN EXPERIMENTS"

Prior to the postulation of the organization hypothesis of brain differentiation by Charles Phoenix, Robert Goy, Arnold Gerall, and William Young in 1959,[38] scientists believed that hormones exerted a *temporary* effect on the (sexual) behavior of adult animals. Convinced that pre- and early postnatal exposure to androgen had permanent effects on the brain and thus on adult sexual behavior, Phoenix and his colleagues carried out experiments in which they administered prenatal testosterone propionate (TP) to pregnant guinea pigs. The "female" guinea pigs (some of whom were physically virilized) whose mothers had been exposed to high levels of androgen expressed more mounting behavior than "normal" females but less than "normal" males, whereas those whose mothers had not been administered TP displayed what was taken by the research team to be typical female sexual behavior (i.e., lordosis—arching the back, being receptive to mounting, etc.). Consequently, Phoenix and his colleagues concluded that "androgenic substances received prenatally have an organizing [i.e., structural and permanent] action on the tissues mediating mating behavior in the sense of altering permanently the responses females normally give as adults, resulting in masculine mating behavior."[39] These findings, they argued, dovetailed with those of embryologist and endocrinologist Alfred Jost whose experiments on rabbits at the Collège de France led him to proclaim that male characteristics in the fetus are engendered by the testicular hormones testosterone and anti-Müllerian hormone, respectively responsible for the virilization of the Wolffian ducts, urogenital sinus, and external genitalia, and for the regression of Müllerian ducts. In the absence or inactivity of these hormones, the fetus becomes phenotypically female. Money later refers to this as "the principle of Eve first, then Adam"[40] and argues that it "applies not only to the hormonal control of the differentiation of the genital anatomy as dimorphic" but also "to the dimorphic differentiation of the sexual centers and pathways of the brain that have a part to play in the sexual and erotic functioning of the genital organs,"[41] thus implying an analogous relation between intersexuality and behavioral forms of "gender-transposition."

In 1965, drawing on the work of Phoenix and his colleagues, Money extended brain organization theory to humans[42] arguing that "evidence for a direct neural-organizing effect of the sex hormones on sexual behavior comes from experiments on the lower species."[43] Applying this insight to "gender identity disorder in children"—and all the while acknowledging that evidence of genetic predispositions or vulnerabilities to gender cross-coding is inconclusive—Money thus turned to experiments on female sheep who had been exposed to androgens and who reportedly displayed masculine behavior that is taken to infer masculine brain organization. Here, as elsewhere Money is careful to remind us that "humans are more complex than sheep,"[44] and that "clinical studies support the hypothesis that there is in human prenatal development a sex-hormonal effect on sexual brain differentiation, but that it does not have a hormonal-robot effect of the type described for sheep and other subprimate[45] mammals."[46] In other words, Money favored what has become known as an organizational-activational model of sexual development in which hormones, acting at critical stages in early development, organize brain structures and/or functions—and Money is extremely vague on the details of this—which are then "activated" by hormonal and environmental influences in postnatal life.[47] On this model, prenatal exposure to sex hormones in humans does not determine G-I/R, but rather, as we saw in chapter 4, lowers one's threshold for some kinds of behaviors and/or responses such that, for example, adolescent boys tend to be "more readily aroused . . . [by] sexy pin-up pictures than [are] . . . adolescent girl[s]."[48] There will, however, be ontogenetic differences between boys as a result of their exposure to different postnatal "somatophysiopsychosocial" factors.

The fact that human animals are not reducible to their nonhuman animal counterparts presented something of a problem for Money and his colleagues, as did the legal and moral prohibition against carrying out experiments on human fetuses. In order to demonstrate that prenatal sex hormone exposure in animals has comparable effects on humans, Money and his coresearchers thus turned to "nature's own experiments,"[49] to "the various conditions of birth defect of the sex organs"[50] that, as Jost had (allegedly) shown, were the result of prenatal hormonal (particularly androgenic) organizing action. The test groups on whom the studies were carried out—and whom Money, Ehrhardt, and Masica describe as "analogues" that "simulate [the animal] experiments"[51]—consisted of "gonadal males whose bodies respond only partially to androgen . . . ; gonadal females whose own adrenal cortices androgenized them . . . ; and gonadal females born during the 1950s who were androgenized by the synthetic progestins given their mothers to prevent miscarriage."[52]

In their work with females from the second of these groups—that is, girls with congenital adrenal hyperplasia—Money and his colleagues argue that what most differentiates them from "normal" (i.e., nonandrogenized) girls is that they "were consistently, even boastfully, tomboys throughout childhood."[53] In other words, the girls reportedly[54] liked boisterous physical activity, regularly joined in rough group games with boys, were competitive, showed a preference for "cowboy gear, toy cars, and guns"[55] over dolls, and jeans over frills, and had ambitions that centered on future careers rather than solely on marriage and motherhood. Money, Evers, and Ehrhardt also claimed that the sexual behavior of androgenized girls tended to be different from those of the girls in their control group in as much as the former were more likely to masturbate in the absence of a partner than the latter, and their erotic arousal was less "sentimental," less the sort of "romantic longing for the loved one alone . . . which will, in his absence, require waiting for his return," that is "more typical of the normal female."[56] However, the researchers take pains to note that there was nothing to suggest that the androgenized girls were more likely to become lesbians than their "normal" counterparts. In other words, despite their alleged "tomboyishness," the androgenized girls' sexual orientation is nevertheless deemed "feminine" because it is orientated toward males. This interpretation is typical of brain organization theory (and of the animal research that Money draws on), which, as Rebecca Jordan-Young notes, "always employs a gynephile [toward women]/androphile [toward men] frame for sexual orientation."[57] Within this logic, one can only ever be "feminine" or "masculine," androphilic or gynephilic (rather than hetero or homo), never both and/or neither.[58]

What is apparent in the work by Money and his colleagues that draws on animal studies and studies of androgenized girls is that brain organization theory (inferentially) connects—albeit in complex, contradictory, and often incredibly vague ways—"brain sex" to sexual orientation and thus to G-I/R (since the former is an aspect of the latter). Brain organization theory, as Jordan-Young so nicely puts it,

> rests on a very simple idea: the brain is a sort of accessory reproductive organ. Males and females don't just need different genitals in order to have sex, or different gonads that make the eggs and sperm necessary for conception. Males and females also need different brains so they are predisposed to complementary sexual desires and behaviors that lead to reproduction.[59]

Jordan-Young's critical reading of brain organization theory's "brain," as a sort of "accessory reproductive organ," is illustrated in the most literal way by the following quote from Money, which, unlike Jordan-Young's observation, is totally lacking in irony. Money writes:

> The data of sexology are generated as much between the ears as between the groins—mind-sex and body-sex. The brain is the organ where both conjunct, where eroto-sexual ideation, imagery, and behavior form a union with physiology. The penis and the vulva are very long organs! They quite literally reach up to the brain.[60]

Brain organization theory, then, is fundamentally a story about sexuality. Moreover, endocrinologist Jacob van der Werff ten Bosch[61] claims that brain organization theory subtended (and was thus shaped by) a preoccupation with (the problem of) homosexuality, and the desire to provide an answer to the question of what causes it. While I do not disagree with this claim, I wonder whether the preoccupation he identifies is more properly with "gender nonconformity" more generally, and in particular with the "uncertainty" that it (in its various forms) introduces into a "system" that is held to be naturally dimorphic, and fundamentally heterosexual (i.e., about pair bonding rather than about individuals). A deconstructive or Butlerian reading of this model would suggest, then, that it is haunted by that which is internal to it but abjected by it. This contention is nicely substantiated by Money's coining of the term "homosexology" to refer to the scientific study of the "biology" of homosexuality, which he opposes to purely constructionist accounts that focus on choice and which he calls "spookological." Homosexology, he writes, "is not a science of spooks."[62]

*

In order to explore more fully the ways in which brain organization research on animals, the notion of gender cross-coding, and the preoccupation with gender nonconformity and its destabilizing effects come together in Money's work, I want now to turn to a mind-boggling article he coauthored with John G. Brennan entitled "Heterosexual vs. Homosexual Attitudes: Male Partners' Perception of the Feminine Image of Male Transexuals" (1970).

## FUCKABLE BIRDS? OR THE MATTER OF AB/NORMAL DESIRE

Like much of Money's published work on transsexualism, this paper deals somewhat tangentially with transsexuals, its primary focus being the men who are in relationships with trans women and the question of whether or not such men should be considered homosexual. To the extent that the article does discuss trans women, the claims made are largely conjectural, as well as being incredibly sexist, misogynistic, and transphobic. Consequently, my initial feeling on first reading the paper was to dismiss it out of hand. However, subsequent readings have led me to regard the paper as providing a key to the motivations that drive Money's work and that shape his view of being-in-the-world (as essentially dimorphic). In short, to me, the paper poignantly illustrates what I will articulate in detail in the last section of this chapter as Money's heteronormative orientation(s).

The paper begins with a summary of popular ideas about what constitutes heterosexuality and homosexuality and highlights the central (but problematic) role accorded genitalia in such conceptions. Money and Brennan write:

> If, prior to surgical sex reassignment and while he has a penis, a male trans[s]exual [i.e., an MTF] has an erotic relationship with a man, then it is a case of two people with a penis having sex together. . . . Common sense defines a male homosexual relationship as that of two people with a penis having sex together. The criterion is behavioral and etymologically literal, according to the evidence of common sense, which does not define as homosexual a relationship in which a sex-reassigned male transexual [an MTF] with a vagina is having sex with an anatomically normal male partner with a penis. Such a relationship is, again by the canon of common sense, heterosexual.[63]

But despite the fact that in the dominant imaginary a penis + a vagina = heterosexuality, Money and Brennan claim that "nonetheless, one wonders about the partner with the penis in . . . a relationship [between an anatomically "normal" man and a trans woman], and wants to know whether he has perhaps some degree of so-called homosexual personality."[64] Exactly why one might wonder about the "sexual personality" of people who have penises and are attracted to trans women is never explained by the authors, and this is both telling and constitutive of such wondering as "normal." I want to suggest that the desire to determine the "sexual personality" of such men—a desire that Money and Brennan universalize (and naturalize) but which is, in fact, their own—is informed by a need to confirm, both to themselves and to others, that SRS will not produce homosexuals. While Money and Brennan never explicitly state this, their thesis is that SRS produces genital effects that visually confirm what was always already the case, that is, the androphilic orientation of the trans women interviewed and the gynephilic orientation of their male partners. SRS is thus constituted as justifiable insofar as it makes things appear "as they should (be)."

In order to determine the gynephilic orientation (the male brain organization and the absence of gender cross-coding) of the male partners interviewed, Money and Brennan develop a series of questions that focus on the circumstances in which the couple met, the relative age of the male partners, their education and occupation, the duration of the relationship, previous contact with "gay" people, history of sexual experience, and "the image as determinant of masculine response."[65] What the authors allegedly find is that most of the men did not have a history of frequenting places where transsexuality and/or homosexuality is common and explicit. All of the men had had previous heterosexual relationships, and while two of the seven had had sexual encounters with men, one was a one-off experience that occurred in adolescence, and the other only happened under the influence of alcohol. The majority of men were found to be younger than their MTF partners, and in terms of education and occupation Money and Brennan conclude that "all but one may be considered occupational transients or underachievers."[66]

While the picture the reader is offered of the men interviewed is one of relative "normalcy," it nevertheless gives the impression that such men are far from the world's most eligible bachelors. It is my contention that this depiction is neither purely descriptive/objective nor accidental. In constructing the men interviewed as at once "normal" (i.e., not homosexual or paraphilic) and less than ideal, Money and Brennan's study constitutes both the men's desire for transsexual women and SRS as (relatively) palatable insofar as the latter does not produce (or enable) homosexuality, but nor does it threaten the continuation of the species by diverting the cream of the male crop away from reproductive heterosexuality. In fact, as Money and Brennan tell it, establishing a relationship with their trans partners helped the men interviewed to "settle down" and become gainfully employed, to become productive—even if not reproductive—citizens. They write: "it appears that the partners of transsexuals needed the stabilizing influence of a continuing intimate relationship and that they did, in fact, respond favorably as is often the case in ordinary heterosexual marriage."[67] This construction of the male partners of trans women as at once "normal" but less than ideal clearly illustrates Ruth Doell's claim that "power hierarchies are maintained in myriad ways but most important among these is the ranking of individuals on the basis of alleged inherent differences. . . . The ranking of homosexuality as inferior to heterosexuality [and, I would add, of "imperfect" heterosexuality as inferior to "ideal" heterosexuality] is seen as necessary in order to maintain the superior status of ["normal"] male heterosexuals which would be threatened by the recognition that [identity] categories . . . are socially constructed"[68] and the effect of situated perceptual practices.

If Money and Brennan's characterization of the men as heterosexual is to be convincing, it is, of course, necessary to demonstrate that the women to whom they are attracted are androphilic in their orientation and brain make-up (even if they are not phenotypically female). For Money (as we have seen in his reference to animal studies and his work with intersex patients), androphilic orientation is most apparent in erotic life. However, in this particular study, "freedom of sexual play and positioning" is seen by Money and his coresearcher as restricted by the fact that only two of the trans women had undergone vaginoplasty, and in both cases the results were not entirely satisfactory and further surgical intervention was required. Consequently, they write, "fellatio and anal intercourse [are] given a more prominent role than usual among ordinary heterosexual couples—to the dissatisfaction of the transsexuals themselves, as well as their partners."[69] Given the potentially dubious performance of these couples in the bedroom, Money and Brennan thus search for other signs or expressions of feminine brain organization. What they see—and I will come back to the question of vision in the final section of the chapter—is what many contemporary readers might regard as the most blatantly sexist stereotypes of femininity imaginable.

As I mentioned earlier, of the men interviewed all but one was younger than his MTF partner. This age differential (which is reversed in the majority of heterosexual relationships) is explained by Money and Brennan not in terms of anything particular to the desire of the men interviewed, but rather, as an effect of the fact that "following the tradition of feminine wiles, [trans women] tend socially to declare their age as lower than it is."[70] While the authors acknowledge that there may be pragmatic reasons for this—for example, to avoid having to account for the period in which the individual was transitioning—they nevertheless reinforce the image of the trans woman as duplicitous and mercenary when they write, without offering any evidence to support their claim, that "those who marry an older man may do so quite explicitly for reasons of financial security."[71] Of the seven aspects of the interviewees' lives on which Money and Brennan's research focuses, six are explicitly directed at the male partners of the trans women. One, however, is noticeably different in that it is concerned with the behavior of the trans women themselves. While this may appear strange given that the paper sets out to address the "issue of whether the [male] partners should be considered homosexual or not,"[72] it is less so if we remember that for Money sexology's primary focus is the couple, and not the individual.

The section of the paper entitled "'Strings Attached'—Contingent and Instrumental Demands on Partner"—which interestingly is given considerably more space than any of the sections that focus on the various aspects of the men's lives—begins with the claim that in each of the partnerships studied, "the transsexual's womanly and wifely loyalty [is] contingent on some instrumental role which the husband was and is expected to fill."[73] The authors continue by noting, first, that to varying degrees this is true of all human loyalties and relationships, and second, that "the contingent loyalty of a transsexual wife to her husband" parallels that of her non-transsexual counterpart, and thus "is neither characteristic nor unique."[74] Why then, one wonders, do Money and Brennan even bother raising the issue of loyalty as contingent except perhaps to imply that it might be preferable it if weren't? The answer lies in the "impression" that the (allegedly objective/scientific) researchers have that the seven cases studied "indicate what is, indeed, a strong trend among male transsexuals in their womanly role, namely to be devious, demanding and manipulative in their relationships with people on whom they are also dependent"[75] (including, interestingly, their doctors). This statement is qualified by the claim that "the seven cases are not representative of all male[-to-female] transsexuals," which again leaves one wondering why an unsubstantiated impression formed by "observing" a statistically insignificant number of individuals whom one admits are not representative of a group more generally is considered worth mentioning or, perhaps more particularly, publishable. As in the interpretation of the men in the partnerships studied as normal but less than ideal, this depiction of their MTF partners likewise constitutes the latter as at once "typically feminine" and other than ideal inasmuch as their performance of femininity is "excessive." In short, as Money and Brennan see it, trans women's "feminine wiles" are less driven by the sort of romantic desire, the feelings of love, that are allegedly typical of and appropriate to women and more by perfunctory, instrumental, and opportunistic motives. In a statement that beggars belief—in terms of its content, its audacity, and its purely impressionistic status—Money and Brennan write, "Though there is no definite proof at the present time, it is quite likely that one of the characteristics of the transsexual condition in males [MTFs] is impairment of the neuropsychologic mechanism that mediates the experience of falling in love."[76] In other words, while a relationship between a "normal" man and a trans woman is not homosexual as such, neither is it ideally heterosexual.

Given the largely negative behavioral characteristics attributed to trans women by Money and Brennan, one wonders what may have attracted their male partners to them. The answer, the authors contend, can be found in studies undertaken on a variety of bird species. In order to "prove" that the initial response of the male interviewees to their MTF partners is not indicative of homosexuality—that it is "that of a man to a woman,"[77]—Money and Brennan argue that "the answer to this apparent paradox"—that is, the fact that a "normal" man would be attracted to a trans woman, and moreover, that he would choose her over a "normal" woman—lies "in the biology of imagery, and specifically in the power of a given visual image to a specified reaction."[78] As I mentioned earlier, Money and Brennan associate erotic arousal through visual means with masculine brain organization. They write, "men's erotic arousal, including erection of the penis, is eye-sensitive, and can be rather easily triggered by a visual stimulus." This sounds horribly like the (sexist) adage that most ("normal") men will "fuck anything in a skirt." There are, however, limits, claim Money and Brennan, to "how much the visual stimulus may be reduced or expanded before losing its triggering power," and they illustrate this point with the suggestion that if a trans woman dresses "as a man" she is unlikely to attract "normal" men, but "the odds change" when she "dresses as a woman and effectively presents . . . [herself][79] socially as a female."[80] The effectiveness of a particular woman's appearance will vary depending on the ontogenetic predilections of individual males, but phylogenetically, appearance stimulates an erotic response in all individuals with masculine brain organization, and this is apparent, Money and Brennan suggest, if one looks at experiments that have been carried out on turkeys, chickens, and the Brewer's blackbird.

In each of the studies that Money and Brennan cite, the researchers use taxidermic models of female birds in various stages of dismemberment to determine what (visually) evokes erotic arousal, and, more particularly, the level at which male birds lose sexual interest in potential partners. For example, in their 1952 study of the Brewer's blackbird (whose plumage is sex differentiated in color), Howell and Bartholomew write (and this quotation is cited by Money and Brennan):

> To obtain a mating response from the male, a dummy must meet certain minimum conditions. The wings are not necessary. Either a head or a tail must be present, but one or the other may be removed without eliminating the response. Further removal of parts inhibits the mating reaction. . . . The plumage color should be predominantly that of a female, but a female head and neck on an otherwise male-colored dummy may be effective when the male is in a state of high sexual excitement.[81]

Money and Brennan then reiterate an encounter that is horrifying in its implications (although both Money and Brennan and Howell and Bartholomew seem oblivious to the literal and symbolic violence being performed in this incident and in their respective reportings of it). Citing Howell and Bartholomew, Money and Brennan state,

> A female dummy with a male head, neck and breast evoked mounting behavior, but not copulation. The mating responses of the males to this dummy were inhibited by the male characteristics of the head and neck. In fact, one male was so infuriated, that when he "alighted directly on the back of the dummy (and), started to copulate, (he) looked down and seemed suddenly disturbed, hopped into the air, and thereby knocked over the dummy. He looked at it from the side and gave it an aggressive peck on the head."[82]

As these quotations make clear, Money and Brennan perceive (and thus constitute) the MTF transsexual and the dismembered birds as analogous. At the risk of overquoting, I again cite Money and Brennan who write, "like incomplete bird models, the human transsexual male, though an incomplete and impersonating female, obviously projects at least the minimum number of feminine cues needed to attract the erotic attention of a normal male."[83] However, they also suggest that as with the stuffed birds who initially aroused the interest of the male birds, but could not maintain it due to their compromised feminine appearance—and apparently this had nothing to do with the fact that they are dead, and thus presumably lacking the kinesthetic, olfactory, aural "cues" one would expect (or even require) in a partner—"several male partners of the transsexuals have reported hesitancy when trying to have sex relations prior to their partner's feminization."[84]

To me, both the experiments carried out on birds and Money and Brennan's equating of MTFs with dismembered dummies exemplify what Pierre Bourdieu calls "symbolic violence," and engender, as Howell and Bartholomew's study graphically demonstrates, literal, physical violence: female birds are dismembered in order that the male (both the bird and the researcher) can take up (and shore up) the position of subject, of normalcy. Indeed, the latter is incumbent on the former. Likewise, misogynistic/transphobic violence is naturalized through the representation of a male bird's knocking over and pecking his female counterpart as "instinctive," and as not requiring critical comment. As I mentioned earlier, Money claims that prenatal exposure to hormones in humans does not (unlike in subprimate species) produce robotic behavioral effects, thus one could seemingly not justify transphobic violence in humans by claiming that such behavior is instinctive. However, insofar as the proposition that phylogeny precedes and lays the foundation for ontogenetic differences and that pair-bonding is a "fact of phylogeny . . . [an] inevitable exigenc[y] of human existence"[85] are fundamental tenets of brain organization theory (in humans), then being repulsed by and/or repulsing a potential mate whose gender identity (once revealed) undermines one's G-I/R, remains a relatively "natural" response—albeit a response that one might learn to moderate.

What my reading of "Heterosexual vs. Homosexual Attitudes: Male Partners' Perception of the Feminine Image of Male Transexuals" has shown is that the androcentric bias (the sexism, transphobia, and homophobia) therein (and indeed apparent in Money's work more generally) is the necessary (and generative) effect of a particular way of seeing/knowing, of what, in chapter 1, I referred to as "biological foundationalism." Biological foundationalism, as I argued earlier, is integral to Money's linear model of G-I/R (or mindbrain sex) development in which there is a unidirectional relationship between pre- or postnatal hormone levels and behavior. This claim is supported by Money's confidence that the answer to what engenders transsexualism "will one day be found to lie . . . in neurophysiology as well as neuropsychology."[86] Money's assumption of a direct one-way relationship between pre- and early postnatal brain organization and adult behavior means, as Longino notes, that "the organism is thereby disposed to respond in a range of ways to a range of environmental stimuli. . . . It also implies a willingness to regard humans in a particular way—to see us as produced by factors over which we have no control."[87] This critique of Money's model of G-I/R development is clearly at odds with his characterization of the model as interactionist, but as Doell has argued, and I agree, a developmental model in which complexity and interaction can only be explained as "added to or subtracted from a 'main cause'" or foundational substrata is not "truly interactionist."[88]

## DISORIENTING MONEY'S VISION

Throughout his writings, Money goes to great lengths to stress the "scientific" status of his research (and of sexology more generally) and to distinguish it from "sexosophy," which he characterizes as ideological, philosophical, and literary.[89] There are a range of motivations that one might posit to explain this, but rather than listing those here I want instead simply to suggest that the imperative to identify (and be identified) with science both shaped and was shaped by Money's turn to brain organization theory and the experiments that had begun to be carried out on animals in the 1950s. Interestingly, however, on more than one occasion Money expressed an interest in phenomenology. Indeed, in discussion of a study carried out in 1955 in an article published a decade later, Money states, "I had then, as I do now, a philosophical commitment to the principle of defining gender-role phenomenologically."[90] Given that, as I see it, the fundamental problem with Money's account of G-I/R acquisition is inextricable from his investment in "scientificity"—that is, the principles that "the criteria of classification are stable, all-inclusively exhaustive, and mutually exclusive,"[91] and that the observations on which such classifications are made are neutral and objective or "value-free"[92]—I want to turn now to Maurice Merleau-Ponty's work on the phenomenology of perception.

Given his vehement rejection of the nature/nurture dichotomy, the challenge Money faced throughout his career was to explain how processes presumed to be internal to the body (for example, hormone production and exposure, the development of neural pathways, and so on) and external to it (child-rearing practices, sex-rehearsal play, etc.) interact to produce G-I/R. In this chapter, I have argued that Money largely failed in this endeavor. If, however, rather than taking up the findings of brain organization theorists, Money had turned to the work of phenomenologist Merleau-Ponty (who did, incidentally, engage with animal studies),[93] his response to the challenge may have proven more fruitful. For Merleau-Ponty, being, or, in Money's terms, G-I/R, is both bodily, and indissociable from the world of others in which it occurs. More particularly, the "root" of selfhood and of the self's relations to the world of meaningful things is the moving body. As Merleau-Ponty tells it, "the body" is less a biological object than the vehicle in and through which the self is oriented, and orients the world. What this means is that the self (or, we could say, G-I/R) and the "things" with which one identifies, engages, and so on come to matter (are engendered and made meaningful) in accordance with the paths that are available to us. And of course, these paths, or ways of knowing/seeing/being, of orienting and being oriented that we inherit, are historically and culturally specific. Moreover, the ways in which we inhabit the world become habituated: as Money notes, the more a particular road is taken, the more its orientation (its ways of seeing/knowing/being) becomes "nativized," the more it feels natural and inevitable, the more it feels like an expression of who we are rather than something that makes us be. The paths we take simultaneously align us with particular forms of social organization—for example, "heterosexuality"—and render other possibilities "out of reach," and again, Money's view of sexuality as either androphilic or gynephilic is evidence of this: in Money's model, it is impossible to think/see/live sexuality as both, or other than, androphilic/gynephilic. For Money this form of organization (androphilia/gynephilia) is a natural fact that subtends culturally variant expressions of gender and sexuality, but from a phenomenological perspective, gender and sexuality are an effect of social organization. This is not, however, to suggest that gender and sexuality are ideological constructs that somehow shape (or obscure) consciousness. Rather, gender and sexuality are bodily orientations that become habituated, sedimented, and thus lived as real. Consequently, they cannot simply be relinquished or radically changed at will.

As we know only too well, worlds unfold predominantly along (already given) lines of privilege that are the effect of sedimented histories, of culturally shared repeated and/or habituated ways of seeing/knowing/being. And, as Sara Ahmed notes, "following such lines is 'returned' by reward, status and recognition,"[94] whereas not following them constitutes the lived experience of some modes of bodily-being-in-the-world as "out of place" or, as Money might put it, "transposed." In other words, the tauto-logic that produces the "exclusionary matrix" and is reproduced by it, generatively effects what it claims merely to name thus rendering particular forms of bodily-being-in-the-world structurally "out of reach," "naturally" "wrong" and/or undesirable, and naturalizing this particular (hegemonic) vision. Nowhere is this clearer than in Money's brutal perception of and orientation to transsexualism.

The phenomenological account of orientation as informed by and informing perception provides a counter to the assumption of objectivity that underpins both Money's "observations" (of various "others")/aspirations and those of the researchers whose experiments on animals he accords an evidentiary status. For Merleau-Ponty, visuality is the effect and vehicle of sedimented contextual knowledges, rather than a neutral process that provides access to empirical objects/facts. Roger Lancaster makes a similar claim when, in his critique of "sex" as it is conceived and/or constituted in scientific texts, he states, "physical bodies, like the material world that encloses them, really exist. But . . . nothing is actually self-evident about what will be *seen* as self-evident in the nature of the body. . . . What so often appears self-evident and timeless . . . belongs to history, not to nature."[95] For Lancaster, then, as for Merleau-Ponty, perception[96] is "an embodied social and collective art,"[97] one that constitutes things as similar or different, familiar or strange, worth noting/noticing or not, worth reporting or not. Such "things," argues Lancaster, "are not self-evident in the nature of the world. They depend on what perspective one takes—or refuses to take."[98]

This understanding of perception as always already "of-the-world," always already a "seeing-with" that shapes/orients the seer and the seen, the knower and the known, such that they are never entirely separate, allows us to conceive Money's vision of gender non/conformity (as visibly self-evident), as perspectival, situated and habituated rather than objective and true (or false). It also allows us to consider the vision he offers us as "his own," and, at the same time, as something more and something other than simply individualized sexism, transphobia, homophobia, and so on. As Merleau-Ponty's account of the blind man's stick as an instrument in and through which he habitually perceives (rather than consciously interprets) the world clearly demonstrates, interpretation is inseparable from perception. Or, as Linda Alcoff puts it, "Our experience of habitual perceptions is so attenuated as to skip the stage of conscious interpretation and intent. Indeed, interpretation is the wrong word here: we are simply perceiving."[99] Rather than looking for certainty, then, as Money in the role of scientist surely does, we might instead relearn to look at the world of gender as/in a project without end.[100] Such an approach has the potential to disorient hegemonic accounts of ab/normalcy, and to open up new horizons, rhizomatic paths, alternative cartographies.

## NOTES

1. For a list of those involved in the clinic at its inception, see John Money and Florence Schwartz, "Public Opinion and Social Issues in Transsexualism: A Case Study in Medical Sociology," in *Transsexualism and Sex Reassignment*, ed. Richard Green and Money [1969] (Baltimore: Johns Hopkins University Press, 1975), 267. See chapter 3, note 22, for information about the Erickson Educational Foundation's financial support of the center.

2. Joanne Meyerowitz, *How Sex Changed: A History of Transsexuality in the United States* (Cambridge, MA: Harvard University Press, 2002), 219.

3. The patient was referred by Dr. Harry Benjamin, and the surgery took place in February 1965.

4. Meyerowitz, *How Sex Changed*, 219.

5. See the press release of November 21, 1966, appended to Money and Schwartz, "Public Opinion and Social Issues," 268.

6. Money and Schwartz, "Public Opinion and Social Issues," 263.

7. For example, John Money's writings on transsexualism are not included in *The Transgender Studies Reader*, and of the fifty pieces in the collection, only one (the paper by Judith Butler) discusses Money, and that is in relation to his role in the David Reimer case. See Susan Stryker and Stephen Whittle, *The Transgender Studies Reader* (New York: Routledge, 2006).

8. John Colapinto, *As Nature Made Him: The Boy Who Was Raised As A Girl* (New York: Harper Collins, 2000), 2.

9. John Money, *Gendermaps: Social Constructionism, Feminism, and Sexosophical History* (New York: Continuum, 1995), 96.

10. Money offers this as an alternative to the "commonsense" model that proposes two sex-based developmental paths.

11. John Money and Patricia Tucker, *Sexual Signatures: On Being a Man or a Woman* (Boston: Little, Brown, 1975), 6.

12. Money, *Gendermaps*, 104.

13. Money and Tucker, *Sexual Signatures*, 143.

14. John Money, "The Concept of Gender Identity Disorder in Childhood and Adolescence after 39 Years," *Journal of Sex and Marital Therapy* 20 (1994): 163–77, 170.

15. John Money and Clay Primrose, "Sexual Dimorphism and Dissociation in the Psychology of Male Transsexuals," in *Transsexualism and Sex Reassignment*, ed. Richard Green and Money [1969] (Baltimore: Johns Hopkins University Press, 1975), 115–52, 127.

16. In *Sexual Signatures*, Money and Tucker write: "the transposition made by transsexuals and transvestites involves the complete redirection of their own gender identities; the transposition made by homosexuals and bisexuals involves mainly the part of their gender identity tied into the direction of their erotic interest" (208). However, they also claim that "unlike the transvestite, the complete transsexual does not have two gender identities. His gender identity has swung all the way over against his anatomy" (31).

17. Money, *Gendermaps*, 96.

18. This assumption has been critiqued by a number of writers. See, for example, Iain Morland, "Intersex Treatment and the Promise of Trauma," in *Gender and the Science of Difference: Cultural Politics of Contemporary Science and Medicine*, ed. Jill A. Fisher (New Brunswick, NJ: Rutgers University Press, 2011), 147–63.

19. Richard Green, "Childhood Cross-Gender Identification," in *Transsexualism and Sex Reassignment*, ed. Richard Green and John Money [1969] (Baltimore: Johns Hopkins University Press, 1975), 23–35, 24.

20. Sandy Stone, "The *Empire* Strikes Back: A Posttranssexual Manifesto," in *Body Guards: The Cultural Politics of Gender Ambiguity*, ed. Julia Epstein and Kristina Straub (New York and London: Routledge, 1991), 280–304, 290.

21. Money and Primrose, "Sexual Dimorphism and Dissociation," 116.

22. Other statements to which the respondent is required to answer true or false, include "you feel deeply sorry for a mistreated horse"; "the sight of an unshaven man disgusts you"; and "you would rather study mathematics and science than literature and music."

23. In *Love and Lovesickness: The Science of Sex, Gender Difference, and Pair Bonding* (Baltimore: Johns Hopkins University Press, 1980), Money elaborates a provisional classification of what he sees as the nine parameters of sex-shared/threshold-dimorphic behavior that provide a largely phyletic basis on which (ontogenic and phylogenic) male/female differences are developmentally superimposed. These are general kinesis—activity and expenditure of energy, especially in outdoor, athletic, and team-sport activities; competitive rivalry and assertiveness for higher rank in the dominance hierarchy of childhood; roaming and territory boundary mapping or marking; defense against intruders and predators; guarding and defense of the young; nesting or homemaking; parental care of the young, including doll play; sexual mounting and thrusting versus spreading and containing; erotic dependence on visual stimulus versus tactual stimulus arousal. See *Love and Lovesickness*, 29–32.

24. Money and Primrose, "Sexual Dimorphism and Dissociation," 129. As I explained in chapter 1, Money refers to the latter as the opposite or negative-valence model of G-I/R.

25. Money and Primrose, "Sexual Dimorphism and Dissociation," 115.

26. Money and Primrose, "Sexual Dimorphism and Dissociation," 119.

27. Money and Primrose, "Sexual Dimorphism and Dissociation," 119.

28. John Money and John G. Brennan, "Sexual Dimorphism in the Psychology of Female Transsexuals," in *Transsexualism and Sex Reassignment*, ed. Richard Green and John Money [1969] (Baltimore: Johns Hopkins University Press, 1975), 137–52, 141.

29. The unacknowledged perspectival (and highly conservative) character of such interpretations has been critiqued by a number of writers, including Lesley Rogers and Joan Walsh, "Shortcomings of the Psychomedical Research of John Money and Co-Workers into Sex Differences in Behavior: Social and Political Implications," *Sex Roles* 8 (1982): 269–81, and Marianne Van Den Wijngaard, *Reinventing the Sexes: The Biomedical Construction of Femininity and Masculinity* [1991] (Bloomington: Indiana University Press, 1997).

30. Money and Primrose, "Sexual Dimorphism and Dissociation," 121.

31. Money and Primrose, "Sexual Dimorphism and Dissociation," 121–22.

32. It is still the case that those seeking access to procedures associated with sex reassignment (or gender confirmation) are required to "pass as transsexual." See, for example, Dean Spade, "Mutilating Gender," in *The Transgender Studies Reader*, ed. Susan Stryker and Stephen Whittle (New York: Routledge, 2006), 315–32.

33. Stone, "The *Empire* Strikes Back," 292.

34. Money and Primrose, "Sexual Dimorphism and Dissociation," 123.

35. For a detailed critique of the bias that shapes such perceptions, see Helen E. Longino and Ruth G. Doell, "Body, Bias and Behavior: A Comparative Analysis of Reasoning in Two Areas of Biological Science," *Signs* 9 (1983): 206–27.

36. John Money, "Gender-Transposition Theory and Homosexual Genesis," *Journal of Sex and Marital Therapy* 10 (1984): 75–82, 115.

37. See John Money, "Sex Reassignment as Related to Hermaphroditism and Transsexual-

ism," in *Transsexualism and Sex Reassignment*, ed. Richard Green and John Money [1969] (Baltimore: Johns Hopkins University Press, 1975), 91–113, 111.

38. Phoenix, Goy, Gerall, and Young all worked in the Department of Anatomy at the University of Kansas in the 1950s.

39. Charles H. Phoenix, Roger W. Goy, Arnold A. Gerall, and William C. Young, "Organization Action of Testosterone Propionate on the Tissues Mediating Mating Behaviors in the Female Guinea Pig," *Endocrinology* 65 (1959): 369–82, 379.

40. Basically, the Adam Principle is the name given by Money to the principle of differentiation that requires that "to differentiate a male something must be added." See Money, *Love and Love Sickness*, 5.

41. John Money, *Gay, Straight and In-Between: The Sexology of Erotic Orientation* (New York and Oxford: Oxford University Press, 1988), 18.

42. Jordan-Young argues that Money was the first scientist to make this move. See Rebecca M. Jordan-Young, *Brain Storm: The Flaws in the Science of Sex Differences* (Cambridge, MA: Harvard University Press, 2010), 6.

43. John Money, "The Influence of Hormones on Sexual Behavior," *Annual Review of Medicine* 16 (1965): 67–82, 67.

44. Money, "Concept of Gender Identity Disorder," 72.

45. In *Sexual Signatures*, Money and Tucker suggest that although "primates are our closest relatives, it's still a far cry from monkey to man" (68).

46. Money, "Gender-Transposition Theory," 115. In *Sexual Signatures*, Money and Tucker write, "Nature chained sexual behavior to the sex hormone cycle to insure perpetuation of the species. Up to the primates, the chain is unbreakable" (175).

47. Fausto-Sterling has convincingly argued that the organizational-activational model does a poor job of explaining animal behavior and provides few clues for the study of human behavior. See Anne Fausto-Sterling, "Animal Models for the Development of Human Sexuality," *Journal of Homosexuality* 28 (1995): 217–36.

48. Money, *Gendermaps*, 37.

49. Money, *Gay, Straight and In-Between*, 7.

50. Money, *Gay, Straight and In-Between*, 7.

51. Cited in Jordan-Young, *Brain Storm*, 33.

52. Money and Tucker, *Sexual Signatures*, 68.

53. Money and Tucker, *Sexual Signatures*, 69.

54. Money claims that the information given to him and his coresearchers by the girls (regarding their "tomboyishness") was substantiated by parents and playmates. However, Longino and Doell raise some important questions about the ways in which such "information" is necessarily always already mediated by the girls' status as "intersexed." See "Body, Bias and Behavior," 220–22.

55. Money and Tucker, *Sexual Signatures*, 70.

56. Anke Ehrhardt, K. Evers, and John Money, "Influence of Androgen and Some Aspects of Sexually Dimorphic Behavior in Women with the Late-Treated Androgenital Syndrome," *Johns Hopkins Medical Journal* 123 (1968): 115–22, 120.

57. Jordan-Young, *Brain Storm*, 161.

58. See also Jordan-Young, *Brain Storm*, 119, and Wijngaard, *Reinventing the Sexes*, 41.

59. Jordan-Young, *Brain Storm*, 21.

60. John Money, "Sexosophy: A New Concept," *Journal of Sex Research* 18 (1982): 364–66, 365.

61. Cited in Wijngaard, *Reinventing the Sexes*, 38.

62. John Money, "Sin, Sickness, or Status? Homosexual Gender Identity and Psychoneuroendocrinology," *American Psychologist* 42 (1987): 384–99, 398.

63. John Money and John G. Brennan, "Heterosexual vs. Homosexual Attitudes: Male Partners' Perception of the Feminine Image of Male Transexuals," *Journal of Sex Research* 6 (1970): 193–209, 193.

64. Money and Brennan, "Heterosexual vs. Homosexual Attitudes," 193.

65. Money and Brennan, "Heterosexual vs. Homosexual Attitudes," 203.

66. Money and Brennan, "Heterosexual vs. Homosexual Attitudes," 198.

67. Money and Brennan, "Heterosexual vs. Homosexual Attitudes," 199.

68. Ruth G. Doell, "Sexuality in the Brain," *Journal of Homosexuality* 28 (1995): 345–54, 353.

69. Money and Brennan, "Heterosexual vs. Homosexual Attitudes," 203.

70. Money and Brennan, "Heterosexual vs. Homosexual Attitudes," 198.

71. Money and Brennan, "Heterosexual vs. Homosexual Attitudes," 198.

72. Money and Brennan, "Heterosexual vs. Homosexual Attitudes," 194.

73. Money and Brennan, "Heterosexual vs. Homosexual Attitudes," 200.

74. Money and Brennan, "Heterosexual vs. Homosexual Attitudes," 201.

75. Money and Brennan, "Heterosexual vs. Homosexual Attitudes," 201.

76. Money and Brennan, "Heterosexual vs. Homosexual Attitudes," 202.

77. Money and Brennan, "Heterosexual vs. Homosexual Attitudes," 203.

78. Money and Brennan, "Heterosexual vs. Homosexual Attitudes," 204.

79. Money refers to the MTF transsexual as "he" and as "himself" here.

80. Money and Brennan, "Heterosexual vs. Homosexual Attitudes," 205.

81. Cited in Money and Brennan, "Heterosexual vs. Homosexual Attitudes," 206–7.

82. Money and Brennan, "Heterosexual vs. Homosexual Attitudes," 207.

83. Money and Brennan, "Heterosexual vs. Homosexual Attitudes," 207.

84. Money and Brennan, "Heterosexual vs. Homosexual Attitudes," 207.

85. Money, *Gendermaps*, 82.

86. Money and Primrose, "Sexual Dimorphism and Dissociation," 130–31. This claim is at odds with the picture that Money and Tucker paint in *Sexual Signatures* of (MTF) transsexualism as possibly being the result of the individual's mother giving them the impression, in childhood, that penises are bad things, and of the consequent rejection of their penis and their masculinity. See Money and Tucker, *Sexual Signatures*, 137.

87. Helen E. Longino, "Can There Be a Feminist Science?," *Hypatia* 2 (1987): 51–64, 58.

88. Doell, "Sexuality in the Brain," 350. In *Gendermaps*, Money writes, "In personal development, from conception onward, that which is individually unique is superimposed on that which is universally shared" (36).

89. See, for example, *Gendermaps*, 136.

90. Money, "The Concept of Gender Identity Disorder," 165

91. Money, *Gay, Straight and In-Between*, 86.

92. For an insightful critique of science as a value-free practice, see Helen E. Longino, "Beyond 'Bad Science': Skeptical Reflections on the Value-Freedom of Scientific Inquiry," *Science, Technology and Human Values* 8 (1983): 7–17.

93. See Maurice Merleau-Ponty, *The Structure of Behavior*, trans. A. L. Fisher (Boston: Beacon Press, 1963).

94. Sara Ahmed, *Queer Phenomenology* (Durham, NC: Duke University Press, 2006), 183.

95. Roger Lancaster, *The Trouble with Nature: Sex in Science and Popular Culture* (Berkeley: University of California Press, 2003), 36; 72–73.

96. For a detailed account of perception as both situated and generative, see Nikki Sullivan, "The Somatechnics of Perception and the Matter of the Non/Human: A Critical Response to the New Materialism," *European Journal of Women's Studies* 19 (2012): 299–313.

97. Lancaster, *Trouble with Nature*, 67.

98. Lancaster, *Trouble with Nature*, 67.

99. Linda Alcoff, "Toward a Phenomenology of Racial Embodiment," in *Race*, ed. Robert Bernasconi (Oxford: Blackwell, 2001), 267–83, 276.

100. Merleau-Ponty claims that "philosophy consists in re-learning to look at the world." See Merleau-Ponty, *Phenomenology of Perception*, trans. Colin Smith [1962] (New York and London: Routledge, 2005), xix.

CHAPTER 6

# "Citizen-Paraphiliac": Normophilia and Biophilia in John Money's Sexology

*Lisa Downing*

> Perversions of the sexual instinct were defined according to the criterion, postulated in Natural Law, that the exclusive purpose of the sexual instinct is procreation.
>
> JOHN MONEY, *Vandalized Lovemaps*

> The figure of the citizen-pervert operates . . . as a constant reminder of the limits of the spaces of sexual citizenship; a figure tucked between the rigid notions of public and private, between sin and crime, disrupting, destabilizing, disordering.
>
> DAVID BELL, *"Pleasure and Danger: The Paradoxical Spaces of Sexual Citizenship"*

In chapter 2, I showed how John Money's conception of paraphilia owes a greater debt, both methodologically and ideologically, to nineteenth-century perversion theory and to turn-of-the-century psychoanalysis than Money was willing to acknowledge. In this chapter, I will explore in more detail, and at some length, one particular set of ideas about paraphilia expressed by Money that places his sexology firmly in the tradition of those "stepchildren of Nature," the nineteenth-century sexologists.[1] That is, I will examine the ways in which Money's writing about paraphilia suggests that it is an antisocial and destructive force, a "slow-spreading epidemic,"[2] that—at its limit—poses a threat to the individual and to civilization, "disrupting" and "disordering" both the public and private order, as the epigraph from Bell suggests.

However, in order to pursue this logic, Money needed to separate "bad" sexual abnormality—the clinical diagnostic concept of paraphilia—from the "good" sexual variation that he and his friends advocated and practiced. As Richard Green puts it in his obituary of Money: "John was a libertine. He was an enthusiast of group sex. SSSS [Society for the Scientific Study of Sexuality] meetings were highlighted by evening orgies organized by John and attended by some of sexology's luminaries. He was a gifted participant."[3] The question of how Money justified this contradiction—and to what extent it really *is* a contradiction—will be central to the analysis undertaken here.

As has already been noted in the preceding chapters of this book, Money's position and arguments, whether regarding transsexualism, intersex, or paraphilia, often appear strikingly contradictory, and it is not easy to ascribe any consistent viewpoint to him. In particular, Money's opinions on sexual practices and behavior seem to alternate between a liberal, sometimes libertarian, sex positivism, on the one hand, and a conservative idea of male and female roles and proper sexual expression, on the other. Money strives to hold in tension a sex-positive agenda, encapsulated in what he calls "a pluralistic democracy of sexualism"[4] with an ethical concern about the dangers of sexual desires that stray too far from the prescribed norm. In the epigraph of *Vandalized Lovemaps*—standing as a comment on the project of the whole volume—he writes of his hope for the creation of a world entirely populated by "paraphilia-free" citizens.[5] The conceptualization of paraphilia as a clinical pathology demanding (chemical) cure is, in fact, both the stumbling block to an obviously consistent political position for Money and yet also the necessary gateway to his establishment as a pioneer in the field of treatment for sex offenders.

In order to balance his openness to plural forms of what he would call "sexuo-eroticism" and the contradictory desire to eradicate some forms of deviancy, it was necessary that Money, like the nineteenth-century sexologists discussed in chapter 2, should believe in the diagnostic quality of perversion/paraphilia as a specific form of pathology, a dysphoria-inducing and dangerous mental disorder. That is, he advocated for the definitional difference between the existence of good, healthy, *optional* "kinkiness," on the one hand, and the *compulsive* condition of paraphilia, on the other. The distinction between the two is defined in such a way that we only have Money's word to take for it: the *diagnosis* is precisely what defines as problematic and pathological both the practice and person "suffering from" the paraphilia. Where an unusual sexual practice is "obligatory" to a person's satisfaction, that patient is considered to be fixated on their paraphilia and diagnosable as mentally ill. Money writes: "variations that are not incidental, but absolutely prerequisite to sexuoerotic pairing . . . constitute a paraphilia."[6] It would be difficult to find a more pertinent illustration of Michel Foucault's contention that the medical technology of diagnosis, and its setting in a particular institutional (clinical) site, function to *bring into being* the "medically peculiar" "personage," here the paraphiliac.[7]

While consciously distancing himself from the judgmental rhetoric of nineteenth-century sexologists with regard to perversion, Money nevertheless uses the metaphorical language of disease to demarcate pathological paraphilia, as seen in the following correlative conjunctional structure: "Just as poverty breeds poverty, so also malnutrition breeds malnutrition, cholera breeds cholera, and paraphilia breeds paraphilia."[8] The implicit message is not subtle: paraphilia, like poverty, malnutrition, and cholera can kill. Whereas, as we have seen, the nineteenth-century sexologists blamed degeneration for the perceived outbreak of moral and sexual perversion, Money blames a repressive, Christianity-influenced society. (The Occidentalism of Money's remit and concerns is glaringly obvious.) Yet the rhetoric Money chooses to express the threat he perceives—"paraphilia breeds paraphilia"—is striking. The "breeding" in question here is pointedly metaphorical, as Money's central contention is that it is via upbringing, not inherited genetic traits, that perversion spreads. And, moreover, as I will be arguing in what follows, it is precisely their lack of reproductive aim—the fact that paraphilias do not lead to "breeding"—that makes them an ideological problem.

In *Lovemaps*, Money harbingers that parents are ever more likely, in ever greater number, to "vandalize" their children's lovemaps, owing to their internalization of "sex negative" societal values. And, in 1989, the warning is reiterated:

> In each generation of population increase there are more children than parents who are exposed to sex negation. Consequently, with each succeeding generation, there is an exponential increase in the prevalence of dysfunctions contingent on ideological sex negation. In public health, the short-term defense against an epidemic is containment, but the long-term defense is prevention. Short-term containment of paraphilias in the population may possibly be achieved, at least in part, by criminalizing them, and incarcerating and executing offenders in a Hitler-type holocaust. By contrast, long-term prevention of paraphilias will be achieved only by biomedicalizing them.[9]

In terms of the logic and lexical field employed, Money's statement perfectly echoes the nineteenth-century theory he seeks to repudiate. "We stand now in the midst of a severe mental epidemic; of a sort of black death,"[10] Max Nordau wrote in *Degeneration* in 1892 about the perceived moral and sexual perversity of his age. Despite his best intentions, Money's attempts to distance his "progressive" science from nineteenth-century ideology serves only to reveal his disavowed debt to it. Money is careful here to ascribe what he sees as unethical or unjust treatment to the institutions of law and judiciary: criminalization, incarceration, and the death penalty are described as the equivalent of a "Hitler-type holocaust" in an example of unforgiveable hyperbole. He does this in order to allow *medicalization*, which euthanizes the paraphilia rather than the paraphiliac, to emerge as the humane alternative. And, ironically, given his avowedly atheistic and antireligion stance, his rhetoric irresistibly recalls the Christian ethic of hating the sin while loving the sinner.

Where Money's approach to paraphilia may seem at best pathologizing and at worst violent in its desire to eradicate "social diversity and individual eccentricity" from the array of sociosexual possibility,[11] it would be a mistake to downplay the fact that certain of Money's paraphilic patients *actively sought out* treatment for their paraphilic desires and reported great benefits from their treatment with him.[12] Nor should we ignore the fact that the search for an effective way of managing violent sex offenders was—and is—a worthy pursuit. However, Money's insistence on the danger of paraphilia goes much further than the aim of assisting rapists and lust murderers to wish no longer to do harm in the service of their sexuality. He writes:

> The goal is to discover the extent to which sex offenders may have an option to self-govern their sexual behavior, and thus to retrieve their human sexual rights to the fullest possible extent. The benefits of sexual research success will be not restricted, however. *They will be extended to others with a kinky sexual fixation or paraphilia that does not offend the law, but offends only themselves or their partners. On the basis of their informed consent, they will be entitled to receive help toward attaining an alternative to paraphilia.*[13]

The question of *why* paraphilias that harm no nonconsenting person should be eradicated along with the criminal ones goes problematically unchallenged. Money states that such desires may "offend" the paraphiliacs themselves, or their partners, without questioning *why*, and in *whose interests*, the paraphiliac may have internalized the notion that his or her desire is "offensive." The flaw in this thinking lies in ignoring the idea that a reformation of culture such that nonnormative—nonreproductive—sexualities would no longer be stigmatized may entirely remove "offense" or dysphoria from the person experiencing such a desire.[14] (Paradoxically, while not asking where the aversion to the idea of paraphilias comes from, or whether it should be supported or challenged, Money admits to "the extraordinary resistance that a paraphile encounters within himself/herself when confronted with the condemnation of other people who are intolerant of his/her difference."[15]) The modification of paraphilic sexuality in the present, and the utopian drive to "liberate" the parents of the future so that they have a more healthy (in Money's understanding of the term) attitude toward their children's sexual development, are dressed in the humanistic, liberal discourse of "rights." By using this argument and language, Money suggests that the drive to eradicate paraphilia is in fact *a part of* liberal rights discourse, not, as it may at first appear, a conservative violation of it. What Money does not consider is the idea that liberal rights discourse *may itself* be a limiting recourse for those who disagree with his (and society's) version of ideal sexuality, or who oppose—or are liminal to—the ideal of universal humanity it presupposes.

I would also posit, to return pointedly to what will be my central argument in this chapter, that, if no nonconsensual harm to others issues from a given paraphilia, one of the only logical objections to that paraphilia would be that it infringes or thwarts the presumed and prescribed aim of *sexuality itself*, as sexuality is understood as a discursive-ideological field. And I would argue that this understanding has changed relatively little between the heyday of the doctrine of Natural Law, quoted in the first epigraph of this chapter, from which Money nominally distances himself, and the late twentieth-century sexual medicine he practiced: the idea that "the exclusive purpose of the sexual instinct is procreation." When Money writes of "paraphilia breeding paraphilia," his figurative language fails to state what is really at stake here: that paraphilia, as nonreproductive sexuality, *refuses to breed*—and that the quality of the threat it is perceived to pose lies precisely in this fact.

No place is opened up in Money's system for the idea of the "citizen-paraphiliac," to borrow from and tweak David Bell's term, as cited in the second epigraph.[16] This is because such an idea would require a radical reimagining of culture, rather than a drive to "correct" errant desire in an attempt to deliver "full humanity" to "paraphiliacs"; that is to make them fit more neatly with the norms of this culture and its understanding of "humanity." In what follows, I will argue that for Money, as for the nineteenth-century sexologists, "paraphilia" qua diagnostic entity is understood, at the unconscious if not the conscious level of the sexological system, to lead to social death. Nonreproductivity, rather than "deviant sexual practices," is thus the enemy of the sexological project in general, and of Money's work in particular.

## NORMOPHILIA/PARAPHILIA; BIOPHILIA/ NECROPHILIA: ANTIPODEAN TENDENCIES IN PSYCHOLOGY AND FEMINISM

In the opening page of *Lovemaps*, Money writes: "Lovemaps! They're as common as faces, bodies, brains. Each of us has one. Without it there would be no falling in love, no mating, and no breeding of the species."[17] In this he makes very clear the positivistic premise underlying his sexology: the purpose and outcome of what the psychoanalysts would call "desire" or "libido" is—at base—reproduction. As early as 1957 in *The Psychologic Study of Man*, Money writes that one of the "inevitables of being human" is that "a person is at birth equipped eventually to be either a mother or a father."[18] The idea that the sexual instinct is programmatically and biologically in the service of species preservation is an idea with a long ideological and scientific history, as I explored in chapter 2. It is also an idea that is somewhat at odds with at least some of John Money's writing, since his avowed adherence to sex-positive politics—which incorporates nonjudgmental attitudes toward homosexuality, bisexuality, and nonreproductive behaviors—would logically preclude a *wholly* reproduction-orientated sexosophy.

The notion that the human being is instinctually inclined or predetermined toward the continuation of life is an idea that found precise expression in the 1970s and 1980s in the psychological and biological sciences in the concept of "biophilia." Psychoanalyst and humanistic philosopher Erich Fromm argued in *The Anatomy of Human Destructiveness* (1973), borrowing from and adopting Freud's model of Eros and Thanatos, that human beings have a propensity to fit either a *biophilic* personality type or a *necrophilic* one. The term "necrophilia" is used here, not in the sense of a clearly delineated sexual perversion or paraphilia, though the label is obviously borrowed from the lexicon of nineteenth-century alienism,[19] but as a character type. "Biophilia" describes "the passionate love of life and all that is alive; it is the wish to further growth,"[20] while "necrophilia" delineates "*the passion to transform that which is alive into something unalive; . . . the exclusive interest in all that is purely mechanical. It is the passion to tear apart living structures.*"[21] It is worth noting that, while these are apparently clinical descriptions, ethical approbation is attributed to the former and negative, pathological overtones to the latter. Biologist Edward O. Wilson, credited with being the first to use the term in the biological-phenotypical context, universalized Fromm's principle of biophilia in a *New York Times* book review in 1979, such that it became "an innate tendency of all humans to focus on life and lifelike processes."[22] He wrote that "modern biology has produced a genuinely new way of looking at the world that is incidentally congenial to the inner direction of biophilia . . . this instinct has the opportunity to be 'aligned with reason.'"[23] This resolutely logical-positivistic and rationality-extolling rhetoric situates Wilson firmly in the philosophical tradition of post-Enlightenment humanism, along with John Money.

However, John Money does not adopt the specific language of "biophilia" to differentiate the outcome of successful lovemap formation ("falling in love, mating, and breeding of the species") from that of the paraphilic, vandalized one (sterile and dangerous sexual activity, compulsorily acted upon). Rather, he coins and popularizes the term "normophilia,"[24] defined in *Lovemaps* in the following way: "Normophilia (adjective, normophilic): a condition of being erotosexually in conformity with the standard as dictated by customary, religious or legal authority."[25] In *Sin, Science and the Sex Police*, written twelve years later, the definition is amplified with the following: "The criteria of normophilia are not absolute but are statistically, transculturally, and ideologically variable. The ideological norm is imposed by those in power, be it parent, peer, clergy, or police."[26] What is extraordinary here—since Money enthusiastically and repeatedly endorses converting a paraphilic lovemap into a normophilic one throughout his opus, and writes in laudatory terms of normophilia—is that this definition focuses on contingent cultural locatedness, and even talks of conformity with the very religious authority Money passionately decries elsewhere. Thus, the status of normophilia is very hard to interpret since Money has written, in passages I have quoted above, of how the social religious standards of his contemporary culture were resolutely "sex negative" and thereby *created paraphilia in the first place.*

What is clear, however, is that Money's "normophilia" fulfills a similar ideological function to "biophilia," while shifting emphasis onto the (social) idea of "the norm" (with its overtones of cultural construction and statistical accountability),[27] moving Money away from biology and toward a social-psychological model. As has been pointed out throughout this book, a feature of Money's entire corpus is a confused and sometimes contradictory attempt to integrate the biological and the environmental, which is undermined by his tendency to lean heavily on one or the other side of the dichotomous pair depending on the argument he wishes to make or the ideology he wishes to oppose. In "normophilia," "norm" is put in place of the "natural instinct for life" (presumed in the biological model of "biophilia"), and the entity that is preserved and promoted under the label "normophilia" is not (just) organic life, but societal harmony—a lack of social disruption, destabilization, and disorder, to return to Bell's terms to describe the citizen pervert's function. One can posit, then, that the coining and use of this term involved a deliberate decision on Money's part to emphasize the social over the organic *in this instance*, since "biophilia" was so available to the psychological and scientific community writing at the time Money produced *Lovemaps* that it would have been an obvious choice to serve Money's aim in describing what paraphilia *isn't*.[28] It is possible that "normophilia," which bespeaks an interest in conformity to *social* life and *social* reproduction, was coined because it suggests etymologically the possibility for conversion that Money hopes to undertake in changing paraphilia—that which runs alongside or beyond normal love—into its sanctioned counterpart. However, "biophilia" is the silent antithesis to "paraphilia" since, as is demonstrable, Money does not, at base, challenge the notion that the purpose of sex is, and should be, reproduction ("mating" and "breeding of the species")—just, in fact, as is taught by the doctrine of "religious authority." Implicitly, then, the couple "normophilia" and "paraphilia" function as perfect counterparts to Fromm's "biophilia" and "necrophilia," each term in the pair bearing the same ideological weight—and having a value judgment ascribed to them.

Another contemporaneous discursive field in which the language of paraphilia was used to suggest death dealing, rather than life giving, was 1970s and 1980s radical feminism. While the pair "biophilia"/"necrophilia" originates with Fromm, as detailed above, this language is also found in the work of Mary Daly, a detractor of sexology and sexual medicine (she describes gynecology as a practice of "Sado-Ritualism"). In Daly's *Gyn/Ecology* (1978), men and male "energy" are described not only as "destructive," but as "necrophilic," in contradistinction to the life-giving energy of women. She criticizes Fromm's description of necrophilia because he makes mention of "hate against women" as merely one "detail on an itemized list" of necrophilous characteristics, rather than acknowledging that, as she claims, "woman hating is at the core of necrophilia."[29] The vocabulary of perversion/paraphilia ("necrophilia," "sadism") that Daly applies to male culture (and by extension to all men) in *Gyn/Ecology* is also applied to trans people. "Transsexualism" is labeled a "necrophilic invasion"[30] of women's space and the trans woman is seen as an antifemale threat for the radical feminist because—as in the case of the nonreproductive paraphiliac for the sexologist—her erotic activity *cannot lead to pregnancy and reproduction*. In short, in Money's favored terms, she cannot "menstruate, gestate, and lactate."[31] Moreover, the postoperative trans body is seen as the product of "male surgical siring," which is understood as an attack on women's innate reproductive power and "an attempt to 'create' without women."[32] Thus, Daly essentializes female *being* as equivalent with the female biological capacity, puts the female life-nurturing body in the place of Nature, and ascribes to the man/trans woman the role of the violent conqueror/usurper of nature and of life.

Radical feminists such as Daly were often the target of Money's disapprobation, as seen when he described their political beliefs as "an attack on masculinity . . . in which male dominance over women is equated not simply with power and aggression but with pornography, harassment and rape."[33] Money also went so far as to critique radical feminists for their efforts to reform rape law such that it would include "the clambering or coerciveness of a male insistent on a genital resolution to a state of sexuoerotic arousal already arrived at by mutual agreement."[34] He considered such expressions of female bodily autonomy "sex-negative." Money's own "feminism"—as far as it went—consisted only in seeking to "liberate" female sexuality from the repression he perceived it to be under from religion, a sex-negative society, *and* from anti-porn radical feminists. That is, he was concerned to enable women to have *more* sex but unconcerned that they be enabled to decline the sex they did not want, or to change their mind about wanting to have sex without having to face coercive consequences.

Although espousing diametrically opposing political aims, then, the strange bedfellows Money and Daly in fact seem to share certain convictions. First, they agree that paraphilia, necrophilia, and sadism are overwhelmingly male characteristics. While Daly refuses to believe that women have anything other than an innate tendency toward life-giving, creative biophilia, Money writes repeatedly of the male's "greater vulnerability" to paraphilic lovemap vandalization, citing the fewer examples of female paraphiliacs seen in his clinic, and states: "it may be that paraphilia is to men what anorgasmia and erotic apathy are to women."[35] Later in the same volume, however, as is all too typical, he contradicts himself somewhat, critiquing "the shibboleth of women as romantic and loving, but not, like men, erotic and sexual," claiming that this belief is held culturally intact only because the medical study of women's sexuality is too often "neglected."[36] It may be that Money's equivocation on the point of women's appetite for (both "normophilic" and "paraphilic") sex acts and fantasies is necessitated by his adherence to the belief that, were social restrictions on (his idea of) "healthy sexuality" to be lifted, women and men would *both* be more sexually expressive. Money also admits that the existence of female paraphilia is a "challenge" to both the *theory* of paraphilia and the *practice* of antiandrogen treatment as a cure. (This is because less impressive results had been produced when the female synthetic hormone Depo-Provera [medroxyprogesterone acetate] was prescribed to the female sex offenders in his clinic than when it was given to the male paraphiliacs.[37]) Also—and tellingly—for both Money *and* for the radical feminists whose politics the sexologist decries as so antithetical to his program of sexual liberation, the biological capacity of sexual activity to lead to impregnation is valued, whereas paraphilias that inhibit this outcome are problematized. In the one case study of a paraphilic female that Money describes in *Lovemaps*—a case of a woman who engaged, from the age of twelve, in genital rubbing against baby boys and male dogs—the case history is framed as "Motherhood Encumbered by a Nepiophilic/Zoophilic Lovemap," and it is presented as a problem that her paraphilic obsession led specifically to her being "immobilized and unable to do her share of the domestic and child-rearing chores. They devolved largely to her husband."[38] The gendered nature of the description of this case study is obvious. Where Money's male patients are framed as sick *individuals*, as we'll see in what follows, his female ones are understood as failed *mothers*. In short, then, sexual energy that is not channeled in the potential service of procreation by both male and female patients is a problem, but when a woman is a paraphiliac, her paraphilia is presumed to be antithetical to the essence of her womanhood (biophilia; creating life and performing motherhood), while masculinity is not assumed to have such a natural intimate relation to life itself in the first place.

## MAKING DEATH LITERAL IN PARAPHILIA

While nevertheless stating explicitly that paraphilias can be "playful and harmless" or "bizarre and deadly,"[39] it is striking that a strategy Money deploys throughout his work is to focus repeatedly on the most extreme kinds of paraphilia, the *literally deadly* ones in order to make general arguments about *all* paraphilias. In *Lovemaps*, two of the five paraphilic case studies he presents are of what he calls "the paraphiliac games of death [which] are asphyxiophilia, or self-strangulation; and autassassinophilia, or staging one's own masochistic . . . murder."[40] And the two book-length paraphilic case studies or "lovemap biographies" that Money coauthored, *The Breathless Orgasm* (1991)[41] and *The Armed Robbery Orgasm* (1993),[42] concern, respectively, an auto-asphyxiophiliac excited by witnessing drownings and stranglings on television, which impelled him to masturbate repeatedly while choking himself with a pair of women's panty hose, and a masochist in a relationship with a "hybristophiliac" woman (someone whose paraphilia is to have a partner who has committed criminal acts). The couple described in *The Armed Robbery Orgasm* engaged in a series of aphrodisiac-motivated crimes before the hybristophiliac-sadist turned herself and her masochistic lover/accomplice in to the police, and both were convicted and incarcerated. When explaining the individual and social danger of paraphilia, then, Money repeatedly uses examples in which criminal danger or material death is both the *aim and outcome* of the sexual act, such that the metaphorical threat to biophilia and normophilia posed by paraphilia is *literalized*. By consistently taking fatal and criminal types of paraphilia as axiomatic to the understanding of paraphilia in general, the cumulative effect of Money's discourse is to suggest that the threat it poses is both very grave and potentially fatal.

Money also uses deathly paraphilias to illustrate the limit of informed sexual consent, a principle he generally vaunts, but which is trumped by the necessity to insist upon the unacceptable nature of paraphilia. In *Lovemaps*, he draws on the principle of "personal sexual inviolacy" that he coined in 1979,[43] which holds that "no one has the right to infringe upon someone else's personal sexual inviolacy by imposing his/her own version of what is or is not erotic and sexual, without the other person's informed consent."[44] However, "sexual democracy," he writes, "is not synonymous with sexual licentiousness whereby anything goes, lust violence and lust murder included."[45] Predictably, the example Money then chooses to use to illustrate the problem of this theory of contractual sexual ethics is that of the "masochist with a paraphilic fantasy of stage-managing his own lust murder [who] meets a sadist with a paraphilic fantasy of lust murdering."[46] He points out that if the "sadist" should be apprehended, he may not be tried for a sexual act, but for murder. Although the partners in the act may have shared a contractual agreement they considered mutually binding, the meaning of that agreement is not legally or socially sanctioned, and its context has no traction within the law.[47]

Money's logic appears somewhat tortuous here. Having lauded the principle of consent, and given us the most extreme example of a consensual pact he can imagine, he then undermines his initial proposition that informed consent is a valid social principle. He does this by stating that, since it would be extremely difficult to prove a killing pact was indeed consensual, one should not worry about respecting the principle of consent after all. Rather, one should remove the likelihood of such a pact ever occurring by removing the *possibility of* the paraphilia. And yet, the tendency toward contradiction that we have been observing between Money's clinical concern about paraphilia's inherent deathly danger and his liberal enthusiasm for kinky experimentation again raises its head, and nowhere so strongly as in the following, extraordinary, extract from a section of *Lovemaps* in which Money describes the "S/M community":

> If at the beginning of a relationship a sadist omits to reveal the full range and culmination of sacrifice involved in his/her sadistic fantasy, then the partner may be destined to a role not of collaboration, but of abused sacrificial victim. There is no hard-edged dividing line between the abusive and the playful paraphilias. Nonetheless many people appear to be anchored on the playful side. . . . The expiatory and sacrificial paraphilias are not invariably malignant. For some they are benign. Statistically, those may rate as abnormal, *but ideologically they are acceptable.*[48]

Contradictions abound here. Money states that some paraphilias are not malignant, but it is, in that case, unclear why he persists in calling such benevolent sexual variation "paraphilia," rather than naming such practices instances of "experimental kinkiness," or similar, since they would not seem to fulfill his criteria for the *mental disorder* that is paraphilia. At moments in his texts, then, "paraphilia" means only the mental disorder that he seeks to eliminate via treatment of existing paraphiles and education of future parents. At others, however, such as in this extract, any nonnormative sexual act (that cannot, in and of itself have "breeding" as an outcome) is identifiable as "paraphilia." Money also admits in this extract that paraphilia is an *ideological* matter—that it is a political category—while elsewhere it is presented as a matter of objective medical, legal, and moral urgency.

There is another glitch in Money's logic here, which can be argued to issue from his insistence on making even the most "perverted" sexuality *look like* the kind of sexuality that is life giving, since paraphilia has to be understood as normophilia that has been hijacked (vandalized). Even the deathly paraphilias are defined, for Money, by an obsession with orgasm. His insistence that the search for orgasm (the event that, in the case of male physiology at least, enables the possibility of "breeding") drives the autassassinophiliac to "stage manage the *possibility*" (rather than the *actuality*) of his/her death at the hands of another person says more about what Money thinks all forms of eroticism are *for* than what a given "autassassinophiliac" may actually *want.* In the case study described in *Lovemaps* of "bisexualism and autassassinophilia in a male hermaphrodite," the patient describes clearly in his own words an un-acted-upon fantasy involving opening fire in public so that the police would shoot him dead: "I'm going to go out there and buy me a gun, and just go somewhere, and start shooting people, and just let, let the police end all this turmoil. I'm scared of that. I really am. I don't really want to hurt nobody but at the same time I want somebody to hurt me."[49] No mention is made of the orgasm that would ensue in the course of the fantasy death. Later, the patient described how he would seek out homosexual encounters in which he "would be beaten up, wounded and possibly left unconscious."[50] Again, orgasm is not mentioned and appears beside the point of the desired encounter (and irrelevant/physiologically unlikely if the practitioner falls unconscious first), yet Money defines autassassinophilia precisely by the fact that "facilitation or attainment of orgasm are responsive to, and dependent upon stage-managing the possibility of one's own masochistic death by murder."

This insistence on orgasm may be explicable in the fact that, for Money's paraphilia-to-normophilia "conversion therapy" to take place, the "death" content of paraphilia (whether literal, as in this case, or figurative as in merely sterile/noncoital) has to be replaced with the "life" content of normophilia (again, revealed as biophilia) through resignification of the genital orgasm. Money is seemingly incapable of imaging the erotic murder fantasy/desire as functioning outside of a corporeal and hydraulic organization proper to "normophilia." The autassassinophiliac wants an orgasm, states Money, not death. The sexologist thereby inscribes an unusual practice in a recognizable, normative epistemological framework—that of "sexuality" (as constituted by the field of sexological knowledge as it is organized at his contemporary moment). The threat of death in paraphilia is used here as an incentive to the would-be libertarian to support the suppression of paraphilia and the conversion of a death-related desire to a life-giving form. In order to argue for the *possibility* of biophilic conversion, the intended outcome of even the most extreme paraphilic sexual activity has to already be in line with the end point of normophilic sexual activity.

*The Breathless Orgasm: A Lovemap Biography of Asphyxiophilia* represents Money's showcase of the successful conversion of a paraphilic lovemap into a normophiliac one. It is, in a sense, the equivalent of the Reimer case in the sphere of paraphilia, with the significant difference that the patient in question endorses to the present day the effectiveness of Money's treatment and the benefits it brought him. In correspondence with the current author in 2011, he writes: "It is a rotten shame that Dr Money is no longer with us. Not only was he the doctor who saved my life with Depo-Provera, we were also very good friends."[51] The book is coauthored by Money, by David Hingsburger, a sex therapist, and by the patient-autobiographer himself (whose pen name is

Gordon Wainwright, while he is named "Nelson Cooper" in the biography). This hybrid work is an impassioned defense of the power of sexological intervention into paraphilia. Tellingly, unlike in the short case study of autassassinophilia presented in *Lovemaps* and discussed above, orgasm *is* central to Nelson's paraphilic practice, as is emphasized in the book's title. The book takes the form of a confessional case study in verse, letters, and prose descriptions of sexual practices and sensations experienced by the author-patient. Nelson Cooper, the self-defined "asphyxiophiliac still living," is the nominal authorial voice, but the text is heavily shot through with a language and rhetoric that regular readers of John Money will easily recognize.

The autobiography describes the onset of Nelson's discovery of the pleasures of self-strangulation, his increasing dependency on the practice, and the fantasies of death by asphyxiation (others' and his own) that he would use as accompaniment to the masturbation in which he indulged up to thirty times per day. Nelson's detailed confession of his masturbatory practices and accompanying fantasies is extremely similar in kind to those produced in the heyday of nineteenth-century sexology, as discussed in chapter 2. Indeed, Money states in the book's prologue that the rarity of information about paraphiliacs' experiences "justifies restoration of the tradition, widely accepted a century ago, of publishing long and thorough case studies."[52] An example of Nelson Cooper's confessional style and content follows:

> I strangled myself in front of an angled mirror, using a nylon pantyhose. I wore a tight pair of men's 100 percent-nylon, see-through bikini underwear and pretended the whole time that a homosexual killer was throttling me. I struggled like mad in front of a mirror which was aimed at my buttocks and my legs. After I choked to the point where my dizziness got too much for me, I broke off the pantyhose, fell to the floor as if I were dead, and immediately masturbated until I climaxed in a super, great orgasm.[53]

This is one of many such passages in the book, in which the degree of pleasure felt in the reported acts and fantasies is in abundant evidence, despite the parallel and juxtaposed rhetoric of desire for treatment and cure.

*The Breathless Orgasm* closes with a letter from Nelson to the American Association of Sex Educators, Counsellors and Therapists extolling the benefits of John Money's chemical cure:

> I have not strangled myself for over a year because, of course, of Depo-Provera. My dosage started at 500mg every 7 days. However, I did have

> some trouble a couple of months ago, and so then my injection went up to 600mg every 7 days. I have remained unstrangled since then.
>
> The fantasies have been changing also, I assume from the therapy sessions I have been getting from Dr Money. I have been told by him the ways that people have sex and that women like to have sex too. He has been giving me a sex education course of sorts which has fed into my brain and my lovemap seems to be straightening out and becoming more normalized as the weeks go by.[54]

It is easy to see that the writer of what the text insists was an "unsolicited" letter, has adopted with an evangelical zeal the discourse of the clinic. Indeed, in the archives of the John Money Collection at the Kinsey Institute in Bloomington, Indiana, in March 2011, I was able to examine the three extant typescripts and proof versions of *The Breathless Orgasm*. The original text is written in Nelson's quite poetic but often ungrammatical and certainly nontechnical language. The version submitted to press has been heavily edited by the coauthors, such that passages are reworded to replace lay terms with Money-coined vocabulary. The final, prepublication proof version is much shorter than the original, and contains far fewer of Nelson's original words than the earlier versions.

Nelson's successful cure is marked by a transformation at the level of sexual fantasy content from death to life, strangulation to coition. Money writes in the concluding "Commentary" section of the book: "only after a long period of combined antiandrogen and counseling therapy did the actresses in Nelson's asphyxiation fantasies change roles and became lovers who engaged in lovemaking and peno-vaginal intercourse."[55] It is noteworthy that not only has Nelson's lovemap become, via the therapeutic process, normophilic rather that paraphilic, but additionally it has been heterosexualized, despite Money's many assurances throughout his corpus that a homosexual or bisexual lovemap is not, in and of itself, "sick." The homosexual killer who sometimes appeared in the asphyxiation fantasies has, tellingly, not been transformed into a homosexual lover. And the act that has been substituted for strangulation is, significantly, the heterosexual act that can lead to reproduction. In short, biophilia has triumphed over sterility in Money's curative process, and the orgasm that seems to be Money's talismatic fetish when discussing any libidinal practice whatsoever has been ideationally reoriented to a socially acceptable (normophilic) and potentially reproductive (biophilic) association.

## ANTISOCIAL QUEER, "SHADOW FEMINISM," AND PARAPHILIC CITIZENSHIP

Death-related sexuality functions rhetorically in Money's texts to mark the extreme limit of human sexual behavior, but also, paradoxically, it is a commonly and consistently used device designed by him to warn almost metonymically against the danger of *all* deviation. The reputed statistical rarity of such paraphilias as "autassassinophilia" is in contradistinction to the number of pages devoted to it in Money's sexological manuals. For this reason, it is strange that there has been so little analysis of these kinds of death-related sexual practices as limit-cases in contemporary critical sexuality studies.

Recent critiques of those social discourses and institutions that ideologically privilege reproductive over nonreproductive forms of relationality, and thereby the life-driven future over a death-haunted present, have been found in the work of American queer scholars such as Judith Jack Halberstam, Lee Edelman, and Lauren Berlant who represent the so-called antisocial turn within queer theory.[56] Antisocial queer posits that the homophobic discourse used against gays and lesbians—that they are unnatural because they do not/cannot engage in reproductive sexual relations—should be embraced from the position of queer politics as a challenge to the normative social order that is predicated on reproductive heterosexuality. Taking up a psychoanalytic language, Edelman urges the queer to occupy "the place of the social order's death drive."[57] Edelman adopts the term "reproductive futurism" to describe the dominant (we might add normophiliac, biophiliac) ideology in which the heterosexual couple is politically enfranchised, the figure of "The Child" used as the guarantor of the future, and the nonnormative sexual subject scapegoated and sacrificed in their name.[58]

Like Edelman's abjected queer subject, the paraphiliacs described in John Money's texts are made discursively to embody the place of culture's death drive. Paraphilia—described sometimes by Money as an "unspeakable monster"[59]—is placed by this delineation on the limit of the social order, since the "monster" is the figure that falls outside of "the ordinary course of nature,"[60] and nature and culture are often tightly interwoven by Money in writing of sexuality. There is no way to imagine a citizenship that "belongs" to the paraphiliac-monster in Money's system, since "paraphiliac" becomes the term that troubles what Money calls "sexual democracy" and Bell and Binnie call "sexual citizenship."[61] Citizenship is always-already sexed and sexual according to Bell and Binnie, since the institution of the family as the unit of society is a heteronormative structure, nominally private, but around which huge amounts of public energy and concern are mobilized in order to police its boundaries, according to what Foucault would call the operations of "bio-politics."[62]

Nick Mulé has argued that "the public/private dyad is . . . most challenged by sexual dissidents, whom Bell describes as 'citizen-perverts.'"[63] The paraphiliac as constructed by Money, through the process of medicalization and through the assumption of his/her mental illness, is not allowed a private or public political agency, even though Money occasionally admits, as in the extract from *Lovemaps* I discussed above concerning S/M communities, that *the construction of paraphilia as socially problematic is itself ideological.* We have seen how, in the case of the nepiophiliac/zoophiliac female patient, whose paraphilic desire made her unable to perform attentive motherhood normally, Money employed a rhetoric of concern about the threat to the order of family life. Citizenship traditionally rests on the perpetuation and reproduction of heteronormative family units in which female labor is both mandatory and invisibilized—and in which *any* female desire that disrupts the smooth course of that labor becomes, in itself, a perversion of the social order.

The feminist implications of the norms of citizenship are addressed by Mulé, who writes:

> Traditional notions of citizenship [are] associated with patriarchy, heterosexism and the nuclear family. . . . In the West, claims to citizenship status are associated with the institutionalization of heterosexism and male privilege. This in turn translates into a particular kind of heterosexuality, that of marriage, and the traditional, middle-class nuclear family—a kind of "good citizenship."[64]

Money seldom discusses "citizenship" in his writings, preferring, as noted, to talk of "democracy," but in one rare instance in which he uses the term—and, indeed, problematizes it—it is precisely to refer to the situation of a married woman's legal name, in a quasi-feminist gesture:

> There's also the situation of a Jane Smith who marries John Jones but continues to use her maiden name. She grew up as Jane Smith, the name is embedded in her sense of herself as a person—why should she throw it away when she marries? . . . The law now holds her responsible as a citizen in her own right. Nevertheless, legally, in most jurisdictions, she is Mrs. Jones.[65]

Money demonstrates an awareness of the flagrant sexism and heteronormativity embedded in the legal discourse of citizenship, but is silent on the matter of its implicit *bio*normativity. Gay assimilationist ambitions in the form of campaigns for the right to marry, adopt children, and become "good citizens" stand as one pole of the queer response to the patriarchal heteronormativity of the discourse of the citizen. On the other side of this lies Edelman's resistant death-driven subject who strategically embraces homophobic discourse for dissident ends,[66] or the proponents of "barebacking subculture" who create alternative forms of kinship through fantasies and practices of HIV infection, as described by Tim Dean.[67]

Similarly, it is possible to argue that second-wave feminist arguments such as Daly's, which essentialize the nurturing, life-giving, *gyn/ergic* nature of women, as discussed above, play into the discourses that make logical the current model of motherhood and female domestic labor. Halberstam has noted this danger in a recent work, and calls on a strategic deployment of the energies of what s/he calls "shadow feminism" as a foil to feminist discourse that can be recuperated for normative ends. S/he writes of

> a shadow feminism which has nestled within more positivist accounts and unraveled their logic from within. This shadow feminism speaks in the language of self-destruction, masochism, an antisocial femininity, and a refusal of the essential bond of mother-daughter that ensures that the daughter inhabits the legacy of the mother and in doing so reproduces her relationship to patriarchal forms of power.[68]

A violent feminist such as Valerie Solanas, with her uncompromising refusal of both the social order and ideas of female passivity, marks for Halberstam a point of resistance that exposes the coerciveness of sexed and gendered citizenship and the production of, in Foucauldian terms, the docile (female) bodies of our social order.

Unlike the gay citizen who can choose strategies of assimilation or queer rebellion, and the feminist who can embrace female nurturing or politically reject it, however, the paraphiliac as constructed in Money's medical discourse has no access to this type of strategy for gaining citizenship, if she or he is to remain a paraphiliac. For there is really no *assimilationist paraphilic strategy*. Access to good citizenship comes only through the conversion of a paraphilic lovemap into a normophiliac one. The paraphiliac who remains a paraphiliac is thus not a citizen, but a figure that shows on what basis, and at what price of conformity, membership within society is bought.

## CONCLUSION

In this chapter, I have shown that in Money's sexology, the extreme forms of sexual practice that get used discursively to warn against the dangers of all paraphilia are those in which the sexual impulse is associated with a destructive fantasy. Neoliberalism's logical limit is the conceptualization of the autassassinophiliac and the asphyxiophiliac, who must not be permitted to attain a pleasurable end, but must convert to the normophilic, biophilic existence that promises (sexual, social, economic, and capitalistic) (re)production. Perversions have, since their creation in the nineteenth century, been conceptualized as threats to a healthy, productive, *re*productive society. When death is literally contained in them, their commentators can make so much more persuasive their argument for the regulation of all sexual behaviors and the institution of that impossibility, a uniformly "normal" world. These observations are analogous with Foucault's reflections on the entry into legal jurisdiction of suicide in the nineteenth century, since "determination to die . . . was one of the first astonishments of a society in which political power had assigned itself the task of administering life."[69] A biophilic social order—an order that practices biopower—constructs desires that are contrary to its legitimating rationale as incomprehensible and "astonishing."

The discourses of sexual science from the nineteenth century to the present that construct perversions or paraphilias continue to insist on their destructive content, on their compulsive element, and on the conditions of obligation for those personages categorized as suffering from them. The extent to which paraphiliacs are in thrall to their deathly passions renders them a danger to themselves and to society, and positions them as other to the sexually functional (life-driven) citizen. The idea of the paraphiliac as abjected noncitizen is a discourse that is internalized by the paraphiliac him- or herself. Nelson Cooper is perhaps the most striking example of this as he actively welcomes and adopts the label "asphyxiophiliac still living" to define himself and retroactively constructs the compulsive nature and meaning of his acts in a generically specific text, and with the aim of proving the effectiveness of the therapeutic cure. Before the "straightening out" of his lovemap, Nelson Cooper was a disallowed citizen. Cure brings with it access to the realm of sociality in the possibility of "peno-vaginal intercourse" and everything that that overdetermined act symbolizes and potentiates in the social realm. The lovemap that includes pleasurable coition belongs to the citizen who can play a role in the structures of society.

This chapter has focused on what may appear to be one of the key unresolvable tensions at the heart of Money's liberalism and sexual libertarianism: the uneasy coexistence between his "progressive" views of sexual pluralism (that is perhaps most passionately expressed in his anticensorship stance with regard to pornography and his advocacy of bisexuality, group sex, and "kinky" experimentation) and his contradictory belief in the existence of "paraphilia" as a discrete pathological entity that constitutes disease and that requires eradication for the good of both the individual and society. We have seen how Money uses a liberal rights discourse, arguing that removing a person's fixated paraphilia constitutes a humane corrective process that enables the citizen to have *more* access to sexual freedom, understood as acceptable sexual variety, *within* the normophilic range. Indeed, Money is keen to assert that "the possible range of acceptable and sexuoerotically arousing activities is greater in a normophilic than in a paraphilic person."[70] Via my discussion of radical antisocial and queer citizenship, I have suggested that the liberal discourse of "freedom" from paraphilia through cure is the gesture Money makes to reduce the suffering of a social monster, while maintaining the pillars of the sociosexual order. The discursive figure of the paraphiliac throughout Money's writing uniquely demonstrates how the distinction between "good" sexuality (normophilia/biophilia) and "bad" sexuality (paraphilia) are both deeply entrenched in, and necessary to the continuation of, a social order predicated on hetero-reproductivity.

## NOTES

1. The term is taken from the title of Harry Oosterhuis's biography of Richard von Krafft-Ebing, *Stepchildren of Nature: Krafft-Ebing, Psychiatry and the Making of Sexual Identity* (Chicago: University of Chicago Press, 2000).

2. John Money and Margaret Lamacz, *Vandalized Lovemaps: Paraphilic Outcomes in Seven Cases of Pediatric Sexology* (Buffalo, NY: Prometheus, 1989), 16.

3. Richard Green, "John Money, Ph.D. (July 8, 1921–July 7, 2006): A Personal Obituary," *Archives of Sexual Behavior* 35 (2006): 629–32, 630.

4. John Money, *Lovemaps: Clinical Concepts of Sexual/Erotic Health and Pathology, Paraphilia, and Gender Transposition in Childhood, Adolescence, and Maturity* [1986] (Buffalo NY: Prometheus, 1988), 4.

5. Money and Lamacz, *Vandalized Lovemaps*, dedication, n.p.

6. Money, *Lovemaps*, 68. In a documentary about paraphilia made in 1995 entitled *Beyond Love*, directed by Peter Boyd Maclean, in which John Money is interviewed, the sexologist restates that "fixation" and the "paraphilic fugue state" characterize the mental disorder he calls paraphilia. He reports having seen patients enter his clinic in this "fugue" state (which is a state of intense, altered consciousness in which ordinary reason is suspended). This claim echoes multiple nineteenth-century ideas, such as the alienists' diagnosis of erotic monomania

(a single-minded delusional abnormality that possessed the subject in one single area of life only, impairing reason) and the widespread end-of-century notion that perversion is related to an epilepsy-like state, epilepsy and sexual perversion both allegedly being symptoms of "degeneration," as discussed earlier, in chapter 2.

7. Michel Foucault, *The Will to Knowledge: The History of Sexuality*, vol. 1 [1976], trans. Robert Hurley (Harmondsworth: Penguin, 1990), 43–44.

8. Money, *Lovemaps*, 2.

9. Money and Lamacz, *Vandalized Lovemaps*, 16.

10. Max Nordau, *Degeneration* [1892], 2nd ed., introduction by George L. Mosse (Lincoln: University of Nebraska Press, 1993), 537.

11. Money, *Lovemaps*, 4.

12. I will discuss, in particular, the case of "Nelson Cooper" of whom this is especially true below.

13. Money, *Lovemaps* 6. My italics.

14. The American Psychiatric Association's Paraphilia Subgroup, under the direction of Ray Blanchard, has revised the diagnosis for the new *DSM-5*, which was published in 2013. The major change involves separating "paraphilias" from "paraphilic disorders," where only the latter are to be diagnosed as mental disorders requiring psychiatric intervention. The basic definition of a "paraphilic disorder" is "a paraphilia that is currently causing distress or impairment to the individual" or causing "harm, or risk of harm, to others." The first part of the diagnosis is similar to the definition of paraphilia in the previous version of the *DSM* where it described sexual behavior causing "clinically significant distress or impairment in social, occupational, or other important areas of functioning." Thus, since "distress" and "impairment of functioning" are how *paraphilia* has long been defined, the diagnostic difference is arguable. Moreover, there is still no attempt to address the *cause* of the named distress or impairment: an innate feature of sexual nonnormativity or a response to rigid societal and medical understandings of what sex should be and should be for? For more on this, see Charles Moser, "Letter to the Editor: Yet Another Paraphilia Definition Fails," *Archives of Sexual Behavior* 40, no. 3 (2011): 483–85.

15. Money, *Lovemaps*, 7.

16. David Bell, "Pleasure and Danger: The Paradoxical Spaces of Sexual Citizenship," *Political Geography* 14 (1995): 139–53, 150–51.

17. Money, *Lovemaps*, xv.

18. John Money, *The Psychologic Study of Man* (Springfield, IL: Charles C Thomas, 1957), 11. He adds "or in the rare case of human hermaphroditism, more a mother than a father or vice versa."

19. "Necrophilia" was coined as a sexological term in a lecture given in 1850 by Belgian alienist Joseph Guislain. See Guislain's *Leçons orales sur les phrénopathies, ou Traité théorique et pratiques des maladies mentales. Cours donné à la clinique des établissements des aliénés à Gand* (Ghent: L. Hebbelynck, 1852), 257. "Alienism" was the name given to the early precursor of psychiatry in Europe. The name suggests the idea that "reason" is something from which the subject can become "alienated."

20. Erich Fromm, *The Anatomy of Human Destructiveness* [1973] (New York: Holt, Rinehart and Winston, 1976), 406.

21. Fromm, *Anatomy of Human Destructiveness*, 332, quoting the definition by H. von Hentig. Fromm's italics.

22. Stephen R. Kellert and Edward O. Wilson, eds., *The Biophilia Hypothesis* (Washington DC: Island Press, 1993), 246.

23. Cited in Kellert and Wilson, *Biophilia Hypothesis*, 246.

24. For more on the logic of "normophilia" and its place in Money's system, see Lisa Downing, "John Money's 'Normophilia': Diagnosing Sexual Normality in Late-Twentieth-Century Anglo-American Sexology," *Psychology and Sexuality*, special issue, "The Natural and the Normal in the History of Sexuality," ed. Peter Cryle and Lisa Downing, 1, no. 3 (2010): 275–87.

25. Money, *Lovemaps*, 266.

26. John Money, *Sin, Science, and the Sex Police: Essays on Sexology and Sexosophy* (Amherst, NY: Prometheus, 1998), 136.

27. For a careful genealogy of the modern use of "normal" from its origins in nineteenth-century mathematics to the application of the "type" in the medical field, see Peter Cryle and Lisa Downing, "Introduction: The Natural and the Normal in the History of Sexuality" (191–99), and Peter Cryle, "The Average and the Normal in Nineteenth-Century French Medical Discourse" (214–25), in *Psychology and Sexuality*, special issue, "The Natural and the Normal in the History of Sexuality," ed. Peter Cryle and Lisa Downing, 1, no. 3 (2010).

28. This decision is perhaps ironic given Money's attack on Foucault and his followers in *Sin, Science, and the Sex Police*, precisely for their understanding that the norms of sexual behavior are a matter of social power relations: "Foucault and his social-science and humanist followers are committed to the principle that the expression of one's sexuality is a moral choice that should not be politically dictated. Hence their stringent resistance to the principle of genetic, hormonal, neuroanatomical, or any other biological determinants of sexuality. They are particularly resistant to the concept of biologically based gender differences, male or female, homosexual or heterosexual. They look the other way if reminded that differences attributable to learning—especially learning that takes place at a critical period of development—become part of the biology of the brain." And: "Foucault's social constructionism has roots in dialectical materialism and in what was called, in the 1920s and 1930s, the sociology of knowledge. It makes tolerable sense when used to explain major social movements and institutions. In the social sciences, however, Foucault's disciples have gone one step further. They use social constructionism to explain all of human nature, including the social construction of paraphilias and the stigmatization of deviance. Social constructionism allows no place for biological and phylogenetic heritage. There are no fixed criteria, only relativistic and arbitrary ones" (107).

29. Mary Daly, *Gyn/Ecology: The Metaethics of Radical Feminism* [1978], reissued with a new introduction (London: Women's Press, 1991), 62.

30. Daly, *Gyn/Ecology*, 71.

31. Money, *Lovemaps*, 27, 117.

32. Daly, *Gyn/Ecology*, 71. This is also the thesis expounded by Janice Raymond in her anti-trans diatribe *The Transsexual Empire* (1979). Raymond was Daly's doctoral student and "The Second Passage" of *Gyn/Ecology* is dedicated to her.

33. Money, *Gendermaps: Social Construction, Feminism, and Sexosophical History* (New York: Continuum, 1995), 108.

34. Money, *Lovemaps*, 154.

35. Money, *Lovemaps*, 30.

36. Money, *Lovemaps*, 145.

37. Money, *Lovemaps*, 144.

38. Money, *Lovemaps*, 254.

39. Money, *Lovemaps*, xviii.

40. Money, *Lovemaps*, 89. The full Money-authored definitions of aspyxiophilia and autassassinophilia are, respectively, "a paraphilia of the sacrificial/expiatory type, in which sexuoerotic arousal and facilitation or attainment of orgasm are responsive to and dependent upon self-strangulation and asphyxiation up to, but not including loss of consciousness" and "a paraphilia of the sacrificial/expiatory type in which sexuoerotic arousal and facilitation or attainment of orgasm are responsive to, and dependent upon stage-managing the possibility of one's own masochistic death by murder (from Greek, *autos*, self + assassin + -philia). The reciprocal paraphilic condition is lust murder or erotophonophilia." (Money, *Lovemaps*, both on 258.)

41. John Money, Gordon Wainwright, and David Hingsburger, *The Breathless Orgasm: A Lovemap Biography of Asphyxiophilia* (Buffalo, NY: Prometheus, 1991). The case described in the full-length book is the same case that Money first describes in *Lovemaps* in the chapter "The Lovemap of Asphyxiophilia."

42. Ronald W. Keyes and John Money, *The Armed Robbery Orgasm: A Lovemap Autobiography of Masochism* (Buffalo, NY: Prometheus, 1993).

43. John Money, "Sexual Dictatorship, Dissidence and Democracy," *International Journal of Medicine and Law* 1 (1979): 11–20.

44. Money, *Lovemaps*, 5.

45. Money, *Lovemaps*, 4.

46. Money, *Lovemaps*, 5.

47. For more on the ethics of the lust-murder pact, see Lisa Downing, "On the Limits of Sexual Ethics: The Phenomenology of Autassassinophilia," *Sexuality and Culture* 8, no. 1 (2004): 3017.

48. Money, *Lovemaps*, 49. My italics.

49. Money, *Lovemaps*, 180.

50. Money, *Lovemaps*, 186.

51. Letter of September 6, 2011.

52. Money et. al., *Breathless Orgasm*, 15.

53. Money et. al., *Breathless Orgasm*, 22.

54. Money et. al., *Breathless Orgasm*, 171–72.

55. Money et. al., *Breathless Orgasm*, 175.

56. Edelman and Halberstam were among the scholars speaking in the panel on "The Antisocial Thesis in Queer Theory," which defined the turn in the field at the MLA Annual Conference held in Washington DC, December 27, 2005.

57. Lee Edelman, *No Future: Queer Theory and the Death Drive* (Durham, NC: Duke University Press: 2004), 3.

58. See also Lauren Berlant, *The Queen of America Goes to Washington City: Essays on Sex*

*and Citizenship* (Durham, NC, and London: Duke University Press, 1997). Here Berlant argues that post-Reaganite America is marked by a discourse of "the national future," obsessed with "pornography, abortion, sexuality, and reproduction; marriage, personal morality and family values" such that "a nation made for adult citizens has been replaced by one imagined for fetuses and children" (1).

59. In John Money, *Reinterpreting the Unspeakable: Human Sexuality 2000; The Complete Interviewer and Clinical Biographer, Exigency Theory, and Sexology for the Thurd Millennium* (New York: Continuum,1994), ix–x, Money discusses the monster as the most apt image to figure the secret of sexuality. In Money et. al., *Breathless Orgasm*, the opening words of the preface are "There was a monster in Nelson Cooper's life. Only later did he discover that it had a name: 'asphyxiophilia.' For most of his youth this monster was unspeakable."

60. Money, *Reinterpreting the Unspeakable*, ix.

61. David Bell and Jon Binnie, *The Sexual Citizen: Queer Politics and Beyond* (Oxford: Polity Press, 2000).

62. Foucault, "Right of Death and Power over Life," in *Will to Knowledge*, 133–59. "Biopolitics" and "biopower" describe the means by which, in the modern period, "methods of power and knowledge assumed responsibility for the life processes and undertook to control and modify them" (142).

63. Nick Mulé, "Equality's Limitations, Liberation's Challenges: Considerations for Queer Movement Strategizing," *Canadian Online Journal of Queer Studies in Education* 2, no. 1, (2006), http://Jqstudies.Library.Utoronto.Ca/Index.Php/Jqstudies/Article/View/3290/1419. Mulé is referring to Bell, "Pleasure and Danger."

64. Mulé, "Equality's Limitations."

65. John Money and Patricia Tucker, *Sexual Signatures: On Being a Man or a Woman* (Boston: Little Brown, 1975), 146–47.

66. See Leo Bersani, "Is the Rectum a Grave?," in *AIDS: Cultural Analysis/Cultural Activism*, ed. Douglas Crimp (Cambridge, MA: MIT Press, 1988), 197–222, 209.

67. Tim Dean, *Unlimited Intimacy: Reflections on the Subculture of Barebacking* (Chicago: University of Chicago Press, 2009). Dean writes: "After two decades of safe-sex education, erotic risk among gay men has become organized and deliberate, not just accidental. The principled abandonment of condoms has led to scenarios of purposeful HIV transmission and, on that basis, to the creation of new sexual identities and communities" (ix).

68. Judith Halberstam, *The Queer Art of Failure* (Durham, NC: Duke University Press, 2011), 124. An earlier attempt to found a feminist politics on perversion is found in Mandy Merck, *Perversions: Deviant Readings* (London: Virago, 1993).

69. Foucault, *Will to Knowledge*, 139.

70. Money, *Sin, Science, and the Sex Police*, 136.

CONCLUSION

# Off the Map

## *Lisa Downing, Iain Morland, and Nikki Sullivan*

We have demonstrated throughout this book that Money was heavily invested in the principle of scientificity and have explored some of the many forms taken by his lifelong attempt to make sexology "genuinely scientific."[1] Interestingly—and perhaps surprisingly, then, given the universalist notion of irreducible gender identity/role (G-I/R) dimensions to which he holds firmly throughout his career—in the introduction to *Gendermaps*, Money argues against "the dogma that science discovers the eternal verities and absolute truths and laws of nature, and that it does so by dispassionately following the process of inductive reasoning," asserting instead that scientific hypotheses are "products of their time and place." While acknowledging the perspectival and situated character of knowledge, Money nevertheless claims that "this relativism applies to particular explanations" but "does not apply to the empirical methodology of science, per se."[2] To many readers, including the authors of this book, the distinction posited here between knowledge and its production may well appear unviable: if, one might ask, knowledge is perspectival and open to ongoing modification, how can the ways of seeing/knowing that shape it (and are shaped by it) not also be? Conversely, if the empirical methodology of science is not relativist "per se," why is it unable to produce or access "eternal verities and absolute truths"? The incompatibility between these claims lies, we contend, at the heart of Money's practice, and, as we have explored in this book, gives rise to a range of contradictions that undermine the viability of a scientific sexology that is "nondiscriminatory, nonjudgmental, and nonstigmatizing," even without cancelling out Money's desire to institute just such a discipline.[3] We think that the resultant tension, between an aspiration to objectivity and a recognition of its impossibility, is symptomatic of a cartographic imperative that subtends (and of course shapes) Money's entire œuvre.

## CARTOGRAPHIES OF FUCKOLOGY

Mapping is common to Money's account of all three diagnostic concepts discussed in this book. Money offers his readers lovemaps, maps of gender development, maps of genital differentiation, maps of language acquisition, and—while we have not addressed the following in the chapters herein—he makes reference also to agemaps, foodmaps, songmaps, soundmaps, speechmaps, and "other maps still to be recognized and empirically demonstrated."[4] Sometimes Money's maps are diagrammatic; he pictures sequential stages of genital development in many of his texts and was photographed holding such a diagram in *Time* magazine in 1973.[5] On other occasions, Money describes his texts themselves as maps. As mentioned in chapter 5, Money and Tucker refer to *Sexual Signatures* as "a road map to show you where you are now as a man or a woman and how you got there. It can help you to keep your bearings in relation to your contemporaries, your parents, your spouse, your children, your grandchildren, society and yourself, and help you steer a steady course through the storms that lie ahead."[6] The purpose of the (metaphor of the) road map, both in *Sexual Signatures* and elsewhere in Money's work, is at once to explain the process(es) of orientation—the reasons why and ways in which we come to take some roads rather than others—and to orient the reader in their relations with others and with a world. In this latter sense, the map is exhortative rather than simply descriptive: as the aviator Beryl Markham once noted, "a map says to you, 'Read me carefully, follow me closely, doubt me not.'"[7] Indeed, Money used his diagrams of genital development (such as figure 2) to explain to intersex patients and (more often) their parents "the rationale of treatment," and thereby to confer "the reassurance and conviction of correctness" about medical intervention.[8] To the same end, parents were sometimes furnished by Money with copies of the diagrams to show to relatives and friends.[9]

Inasmuch as the aim of mapping is, at least conventionally, to "produce a 'correct' relational model of the terrain,"[10] a cartographer is cast traditionally in the dual role of the gallant explorer, the pioneer who "breaks new ground," and the God-like I/eye whose vision is abstracted from, unmediated by, and independent of that which it sees. And this, we suggest, is very much the image Money paints of himself. In regard to the former aspect of the cartographer's role, Money presents his clinical practice as "Pygmalion-like," in the manner

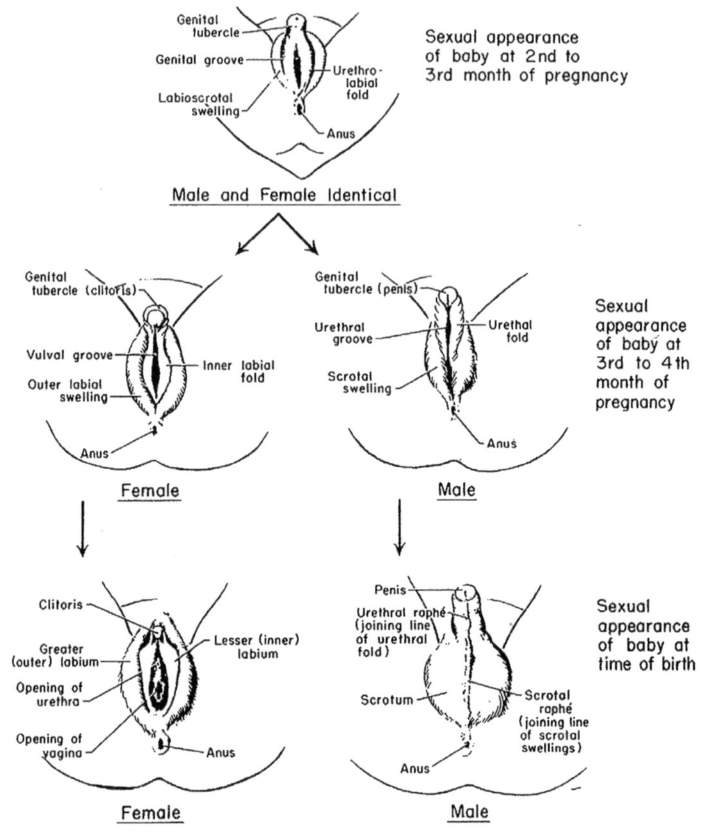


FIGURE 2. Diagram from John Money and Mark Schwartz, "Biosocial Determinants of Gender Identity Differentiation and Development," in *Biological Determinants of Sexual Behaviour*, ed. John B. Hutchison (Chichester: John Wiley, 1978), 769.

discussed in chapter 3; and he writes, in heroic mode, that "impressive as may be the growing body of knowledge on human psychosexual differentiation, no one concerned with research need feel like Alexander, crying for lack of new worlds to conquer."[11] In regard to the latter aspect of the cartographer's role, despite Money's claim to recognize that knowledge is perspectival and open to change, he frames his maps as neutral tracings of what *is*, rather than as necessarily limited visions that are shaped by a desire to conquer and to contain, and which engender the difference(s) they purport merely to describe. This reading is supported by Money's ongoing efforts and exhortations to produce ever more precise map(ping)s of the "nonnormative." For example, in his essay "Biosocial Determinants of Gender Identity Differentiation and Development," Money claims that the task of sexological research "is to identify more and more syndromes of gender identity disorder and impairment. Until this is done, there will continue to be an unspecified multitude of people with problems of gender identity for whom nothing more can be said than that they are idiosyncratically different or deviant—etiology unknown."[12] We might call this a census of the occupants of the duchy of dysfunction. The assumption that mapping is an objective process, which traces what already exists, is also reinforced by Money's interpretation of maps of the past (such as psychoanalysis and nineteenth-century sexology) and map(ping)s produced according to conventions other than his own—for example, the sex/gender distinction that he associates with feminism; the model of "sexual orientation" as something that is, or at least can be, freely chosen; and so on—as inferior, as lacking in the sort of complex detail his maps allegedly provide.

The dual quality of cartography—conquering and uncovering, creating and describing—has been analyzed by scholars in a range of contexts. Feminist philosophers Rosalyn Diprose and Robyn Ferrell explain, "In the metaphor of cartography, to draw a line is to produce a space, and the production of the space effects the line. . . . The lines on the map produce borders beyond which things are seen to be different. Yet the difference of the 'outside' also defines what is 'inside' the border."[13] Judith Butler likewise argues, from a critical psychoanalytic perspective, that the abject(ed) other is both the constitutive outside of the domain of the gendered subject, and that which forever haunts its psychic topology.[14] A map, then, is far from the singular representational object-product of an autonomous and all-seeing I/eye: the map, as a dynamic site of ongoing inter- and intra-action, contains both that which it constitutes as excluded and that which gives it shape in and through its inclusion.

Maps are made by way of montage; they are intertextual assemblages of a plurality of mapping systems.[15] Accordingly, when Money and Ehrhardt claim to assemble "some twenty years of research" in *Man and Woman, Boy and Girl*, they state that the book has been "written at a time when rapid advances in research from various disciplines have opened new vistas against which to reexamine traditional behavioral opinions on masculinity and femininity," yet simultaneously that it "dates back to a monograph on the psychology of hermaphroditism begun in 1949," namely, Money's doctoral dissertation.[16] In the book, the authors try to align observations about female and male behavior that were made on multiple occasions between the 1950s and early 1970s, even though gender norms changed substantially during that period. Such anachronism exemplifies a wider problem in the road maps of gender differentiation that Money used to assess and orient his patients: at the end of the 1970s, after nearly two decades of second-wave feminism and a decade of gay rights activism, Money was still appraising feminine behavior in intersex children as the rehearsal of "fantasies of romance, marriage and motherhood."[17] The territory had changed, so to speak, but the map remained the same. Similarly, in the mid-1980s, when the discourse surrounding the AIDS epidemic had put into deadly crisis the relation between sexual acts and sexual identities, Money differentiated "act and status" by what he called the "Skyscraper Test," which he described as follows:

> One of the versions of this test applies to a person with a homosexual status who is atop the Empire State Building or other high building and is pushed to the edge of the parapet by a gun-toting, crazed sex terrorist with a heterosexual status. Suppose the homosexual is a man and the terrorist a woman who demands that he perform oral sex with her or go over the edge. To save his life, he might do it. If so, he would have performed a heterosexual act, but he would not have changed to have a heterosexual status. The same would apply, vice versa, if the tourist was a straight man and the terrorist a gay man, and so on.[18]

This was a stunningly fatuous way in which to dramatize the relation between risk, sexual act, and sexual identity, at a time when tens of thousands of people had died of AIDS-related diseases, and more urgent than the threat of a "sex terrorist" was the prevailing assumption that the riskiness of a person's sexual acts could be judged by the sexual identity that they claimed.[19] Many of the victims of AIDS had lived in New York City, but in Money's "Skyscraper Test," they were rendered invisible to the cartographer's God-like position atop the Empire State Building.

As the elevated perspective of the "Skyscraper Test" shows, cartography is productive of particular ways of knowing and seeing, and exclusionary of others. In this respect, and in the words of the critical geographer John Pickles, "the map" (as conventionally conceived) "is ob-scene; it directs our attention towards a particular scene, to a particular rendering of a scene; it makes the territory, state, or place stand before us; and it does so by simple devices of intensification—magical tricks and sceno-graphic devices that evoke for us this or that particular world, with all its associations."[20] One such device is the claim to transparency—the idea that the map simply traces what already *is*—which erases or veils over both the "montage origins of all mappings" and the historically and culturally situated quality of the cartographic impulse.[21] This is evident in Money's suggestion that medical information should be "translated" (we might say remapped) into a patient's "conceptual view of the world," thereby becoming "edited, oversimplified, incomplete or metaphorical."[22] Because, by this account, a patient's worldview does not include the ability to discern such redactions, it follows that the explanation provided to the patient, however partial, has the semblance of completeness; in other words, it is a map that conceals the choices made in its construction. Money's diagrams of genital development work in a similar way: they depict intersex genital development only as a stage on the path to sexual unambiguity. Their design reinforces Money's frequent description of intersex genitalia as "unfinished," directing attention toward a particular scene, as Pickles would put it—the scene of "finished," postoperative genitalia that would purportedly resemble those shown in the diagrams.[23] And in a further cartographical trick, the sexually unambiguous genitalia depicted in the diagrams bear none of the signs of surgery likely to accrue during the effort to follow the map toward sexual disambiguation.[24]

Another directive device in Money's writings is the formulation of ever more bizarre terms and (diagnostic) concepts that, like the flags of empire, are planted in (what are constituted as) newly discovered and mastered territories. Without such empirical tools, we are, fears Money, "like explorers lost among a jungle people with a language never before heard, each of us with different lexical ideologies and principles as to how to decipher what the people are saying."[25] Money's sense that communication between individuals, including scientists, would be impossible without a common tongue, is an anxiety to which we will return later in this conclusion. Here, we wish to note the imperialist tone of his remark and its articulation of a familiar colonial trope, whereby a Western explorer's encounter with an off-the-map "jungle people" teaches the explorer something about his own culture. In Money's remark, the lack of a shared language between the explorer and the jungle-dwellers is a trope for the lack of a shared language between explorers. So, the possibility of learning how to speak the "jungle" language of the natives (which might reconfigure the power relations between explorer and natives, self and other) is foreclosed and diverted to a concern with how to speak to other explorers, in order to map the territory on Western terms.

But regardless of the intentions of mapmakers, maps are never simply colonizing and/or normalizing in their effects; our claim in this book is not that Money's map(ping)s repress or distort the real, that without the schemas we have inherited from Money and his ilk, "G-I/R," sexual difference, or sexual pleasure might be lived and experienced more "freely." As we indicated in the book's introduction, our project has been neither to produce alternative, more accurate maps nor to call on others in the field to do so. Rather, our engagement with Money's map(ping)s has been motivated by Marcus Doel's Foucauldian claim that "resistance is assured by the very power of maps."[26] Doel means that insofar as maps are dynamic assemblages, haunted by the constitutive outside that inhabits them, not only maps but also "cartographic reason and the mapping impulse" more generally "are ineluctably caught in the orbit of deconstructive disjointure and dissimilation."[27] Given this, the task of the critical cartographer is to exploit the inherent instability of maps and of the cartographic impulse—the impulse that is so apparent in the work of Money—"in order to make them act otherwise."[28] Therefore, one of our aims in this book has been "to challenge the epistemological myth . . . of the cumulative progress of an objective science always producing better delineations of reality,"[29] by critically mapping the many and varied bodies of thought that have influenced Money's vision—cybernetics, early sexology, brain organization theory, and so on, as well as more pervasive narratives (which are themselves orienting devices) about, for example, the qualities and continuation of the so-called human race. We have shown that the causes and consequences of Money's work resist organization into a sequence of increasingly objective scientific and medical practices; they are, instead, complexly entangled, disorienting, and contradictory, and thereby enabling of certain subject positions and counterdiscourses, such as the identities "intersex," "transgender," and "perv," as well as the possibility of a book called *Fuckology*. Taking seriously Money's observation that the "labels under which we are yclept [classified or typecast] . . . shape our destinies,"[30] and fascinated by the labyrinthine paths he has given us to follow, we have, we hope, vandalized our cartographic inheritance, fucked with the figures of (reproductive) futurity that orient Money's map(ping)s, and, like those stealthy graffiti artists whom we are told haunt our streets when all good citizens are tucked up safely in their proper places, added color and confusion to Money's diagrammatic exhortations.

## REREADING MONEY

As we have argued, Money's rhetoric of mapping implied accuracy, reference, and neutrality, and thereby lent scientific cachet to his work on paraphilia, transsexuality, and intersex. Through his cartographical claims, Money positioned his research as both a value-free enquiry into an objective world, and as the most recent, yet still partial, step toward the creation of a comprehensive map of sex, gender, and sexuality. This strategy enabled Money to deflect criticisms by explaining any inconsistencies in his research as areas of incomplete mapping—such as the brain—rather than as failures to be scientific or nonjudgmental. However, at the same time that the rhetoric of mapping conferred authority, it also signaled equivocation in Money's work. This is because a map is not just a figure for the inscription of its subject matter; in order to function as a guide (to clinical practice, for example), a map must be read. Mapping is therefore a figure for reading, for a map that cannot be read is no map at all. In this section, we demonstrate that even though Money gave prominence in his work to the development of spoken language, most frequently as a synonym for gender development, his comments about the practice of reading provided a more equivocal account of language, wherein Money reflected on the ways in which his writings were read by others.

Money is known as a theorist of sex, gender, and sexuality, but it is equally true to say that he was a theorist of language. Whenever Money discussed gender, he directly or implicitly discussed language, as shown in chapter 4. Apparently unable or unwilling to support his discussions with references to research into language learning, Money's assertions about gender and language were often tautological and speculative—notably in his recurrent suggestions that gender development is at once similar and identical to language development. Yet, despite such synonymy, when Money wrote about reading, he seemed to separate language from gender. In the opening of *Sexual Signatures*, Money and Tucker addressed lay readers directly, asking them, "When you read about a grown man who has become a woman do you wonder if you yourself are a man or a woman? Of course not. You knew that you were a boy, or that you were a girl, long before you learned to read."[31] The purportedly self-evident response of the implied reader to a narrative of transsexuality has, in fact, equivocal consequences. On the one hand, the terse insistence that no reader of a transsexual narrative would question their own sense of being "a man or a woman" presumes a readership composed exclusively of individuals who identify as male or female, and supposes that they will continue to identify as such in the future. It excludes from the book's audience any trans individuals for whom reading about transsexuality could well be transformative, in helping to foster a sense of oneself as a member of a sexual minority, and making visible a path to medical assistance. In this respect, the question and answer posed by the book are conservative of mainstream gender norms.

However, on the other hand, *Sexual Signatures* is subtly subversive, undoing the norms that it advances. The text distinguishes between the acquisition of spoken language, in which readers are presumed to identify from childhood onward as male or female, and the practice of reading. The claim that readers will not question their gender upon reading about transsexuality relies for its self-evidence ("Of course not") on the claim that gender is established before a person learns how to read. This implies that if a person's gender were not established prior to reading about transsexuality, the effect of reading could indeed be to render ambiguous or ambivalent the gender of the reader. In this way, the book suggests that the practice of reading might mobilize atypical gender identifications instead of inhibiting them. Further, it suggests that language is not, after all, an obvious and singular phenomenon to which gender can be straightforwardly compared, but that language is divided between speech and writing. In Money's work, here and more widely, language as it is spoken and heard is the domain of certainty, required for knowledge of oneself as male or female, whereas language as it is written and read is the domain of equivocation. The latter is anchored only by the spoken and heard language that precedes it ("You knew that you were a boy, or that you were a girl, long before you learned to read").

The purpose of our commentary is not to choose one version of language over the other. Rather, from our perspective in the theoretical humanities, the important problem staged by *Sexual Signatures* is that of reading itself. This is to say, the book presents the impossibility of choosing between readings, because the practice of reading, by the text's own account, lacks the unequivocal certainty that characterizes spoken language. Moreover, in problematizing reading, *Sexual Signatures* encapsulates a broader, long-standing dispute over how Money's work should be read, to which Money himself was a contributor. We suggest that by theorizing this dispute as a problem of reading, it is possible to understand his work's contentious qualities without recourse to judgments about Money's character, and without explaining away the shortcomings of his work as failures to make an accurate enough map. On the contrary, because the viability of Money's cartographical project depended on the readability of its maps, the problem of reading is a limit to mapping that cannot itself be mapped. It is what the literary critic J. Hillis Miller, in a different context, has called the "atopical" or "unplaceable place" that interrupts cartography.[32] We submit that the difficulty of reading Money's texts, and ascertaining where they should be placed (for example, in the categories of conservative or subversive, ethical or unethical, science or nonscience), is just such an encounter with the atopical.

The radically irreconcilable range of readerly reactions to Money's work exemplifies the "encounter with the unmappable" described by Miller.[33]

Some readers have called Money's use of language pioneering, lucid, and authoritative:

> He was the first scientist to provide a language to describe the psychological dimensions of human sexual identity; no such language had existed before.[34]
>
> His use of words was masterful.[35]
>
> Money devoted much energy to finding new terms for old, familiar diagnoses. Certainly it made the discussion of sex sound more scientific.[36]

Other readers have described the same body of work in starkly opposing terms:

> The reader must fight through an incredible thicket of contradiction, repetition, dogmatic claims, incomprehensible neologisms, supertechnical language, and an idiosyncratic system of referencing.[37]
>
> [Money's writing has] the propensity simultaneously to inform, shock, entertain, and advise in detached and macabre scientific jargon.[38]
>
> [The] contradictions [between theory and evidence] are couched in so much medical jargon and convoluted reasoning that the reader is swept along bewildered and is hoodwinked into thinking that an integrated theory involving biology and culture is being put forward.[39]

The differences between these reactions cannot be resolved by more reading, even of the nearly two thousand publications to which Money is claimed to have contributed.[40] To do so would be to repeat Money's gesture of trying to make a comprehensive map—in this case, of his own work. One cannot simply cut away the "thickets" of language to clear the ground. Rather, the very irreconcilability of the readings quoted here demonstrates the equivocation of language as it is written and read. The discomfiting quality of Money's term "fuckology" demonstrates this too.

## LANGUAGE AND UNANIMITY

Even though Money implicitly acknowledged the problem of reading in *Sexual Signatures*, a concern for unanimity pervades his work. He tried to foreclose equivocation by ensuring that all readings of his texts corresponded with his own. Money's primary strategy to this end was, of course, the invention of neologisms. His coinages directed others to Money's preferred ways of reading in three respects. First, neologisms let Money present highly conjectural hypotheses as if they were demonstrable phenomena (since, as he once wrote, "an abstract noun . . . becomes father of a thing").[41] Second, they enabled Money to avoid engagement with contemporary critics, by casting his ideas as either entirely different from, or more advanced than, those discussed by others. Third, neologisms facilitated Money's prolificacy, allowing him to complete publications without researching the literature on the topic at hand. An instance of the latter was Money's invention of "paleodigm" in 1989 to name "an ancient example or model of a concept, explanation, instruction, idea, or notion, preserved in the folk wisdom of mottos, proverbs, superstitions, incantations, rhymes, songs, fables, myths, parables, revered writings, sacred books, dramas, and visual emblems."[42] Money's essay on the subject made no reference to scholarship on any of these topics, with Vladimir Propp's *Morphology of the Folktale*—a seminal analysis of common elements in folklore and fairy tales—being a particularly odd omission from a bibliography that nevertheless accommodated nine publications by Money on unrelated matters.

Money's neologisms, then, demarcated a linguistic world composed largely of texts by Money and his coworkers. In the early 1970s, when he reviewed a book about intersex treatment by the British clinicians Christopher Dewhurst and Ronald Gordon, Money remarked that "the time has come for the British to have their own volume [on hermaphroditism], and except in terms of the economics of the book trade, I am not sure why."[43] Accordingly, Money's work was profusely—even predominantly—self-referential, either overtly in citations and quotations or covertly in the verbatim repetition of material between publications. Within that world of language, Money endeavored to assert authority, steering readers toward concepts, values, and conclusions that often had no basis other than their appearance elsewhere in his œuvre. For example, in *Sexual Signatures*, Money quoted the definitions of gender identity and role that he had published three years earlier in *Man and Woman, Boy and Girl*, which were, in turn, closely paraphrased from his 1950s papers.[44] In *Sexual Signatures*, he described these definitions as "official."[45] And, in correspondence with Brenda Love, compiler of the *Encyclopedia of Unusual Sexual Practices* in 1992, Money detailed all those paraphilias he had coined, and which required attribution to him, earning him the following special acknowledgment in Love's book: "Many of the terms in this encyclopedia were coined by Dr. John Money and this special acknowledgement is given to him in gratitude for the many accomplishments he has made in the science of sexology."[46]

In addition to using neologisms to assert authority and measure accomplishment, Money provided an occasional commentary on that very practice. He wrote that there was "a magic about words and a power in technical terms that silences idle curiosity," and reflected that professions "use language to protect power," insofar as "every doctor, lawyer, priest, scientist, and mathematician has to learn, along with the specialized vocabulary needed for precision, a welter of semantic formulas for the sole purpose of maintaining the aura of mystery that shuts out the layman."[47] At times, Money's neologisms had no apparent purpose other than the generation of just such power and mystery, for instance, in the following passage from a 1961 publication:

> The varieties of hermaphroditism are in need of short, one-word names. Such names can be composed if two or three key characteristics of each variety are abbreviated and the syllables added together. Thus, the . . . four varieties of hermaphroditism in genetic males may be named: (1) Gonap (gonadal aplasia); (2) Hypotest (hypoplastic testes); (3) Penutest (penis + uterus + testes); (4) Femtest (feminizing testes). There are four more types of hermaphroditism . . . : (5) Testov (testicular + ovarian tissue); (6) Penov (penis + ovaries); (7) Adrenov (adrenogenital syndrome + ovaries); (8) Techrofem (testes + chromatin-female).[48]

The ambition of this passage is remarkable: Money hoped to be the progenitor of a new vocabulary for intersex diagnoses. However, his angular, pseudo-Slavic portmanteaus were universally ignored.

In the case of gender, contrastingly, Money's word was extensively circulated. Yet, it was also subject to widely divergent readings, as explained in chapter 1. The resultant lack of unanimity caused Money consternation: he described it as "nosological chaos."[49] In formulating the term, Money had tried to craft a memorable and unequivocal definition. Initially, he had defined gender as a role (not an identity), in the following way:

> all those things that a person says or does to disclose himself or herself as having the status of boy or man, girl or woman, respectively. It includes, but is not restricted to, sexuality in the sense of eroticism.[50]

Subsequently, Money adjusted the definition to address the distinction between identity and role, which had been drawn by Robert Stoller in 1964.[51] In its adjusted form, Money's definition ran as follows:

> Everything that a person says and does, to indicate to others or to the self the degree that one is either male, or female, or ambivalent; it includes but is not restricted to sexual arousal and response; gender role is the public expression of gender identity, and gender identity is the private experience of gender role.[52]

Money incorporated his definition of gender into hundreds of publications throughout his career, with very little deviation from the above phrasing. The phrasing is, in fact, unusual in English because of its meter. Both versions quoted above are characterized by heavy use of trochees (an accented syllable followed by an unaccented one) and dactyls (an accented syllable followed by two unaccented ones). So, for instance, "Everything that a person says and does, to indicate" is composed of a dactyl, then four trochees, then another dactyl. The earlier formulation ("all those things that a person says or does to disclose") similarly has a dactyl followed by four trochees, but begins with a spondee, the two stressed syllables of "all those." These metrical choices give Money's definition of gender a sense of tumbling inevitability; its entire first sentence is a cadence, as if it were obvious that a person's speech and behavior could be both itemized and summated.

In the second sentence ("It includes, but is not restricted to, sexuality . . ."), the meter is more mixed. The qualification, "but is not restricted to," is semantically superfluous, yet slows the pace of reading for an impression of thoughtful gravitas. The effect is to make the subject matter, the classification of "eroticism" and "sexual arousal" as aspects of gender, seem carefully considered. The third sentence in the longer version of Money's definition is predominantly dactylic (such as in "public expression of gender identity"); it has the artful bounce of a mnemonic. Finally, the chiasmic arrangement of "gender role" and "gender identity" substitutes grammatical symmetry for reasoned argument. Money never explained exactly why he thought that gender identity and role were the "public expression" and "private experience" of each other, when in practice—for example, for members of sexual minorities—the identity that one holds oneself to be, and the role in which one is cast, can be experienced as profoundly dissimilar. Publicly claiming an identity, especially as a member of a minority, can mean committing to change the role with which the identity has hitherto been associated.

Throughout Money's definition of gender, iambs (an unaccented syllable followed by an accented one) are rare. Everyday spoken English is typically iambic, so Money's unusual meter foregrounds the difference between speech and writing. Curiously, it does so in the opposite way to the rest of his work, eschewing patterns of speech in favor of a meter that is more comfortably read than spoken. The aim was probably to make the definition amenable to written quotation. However, despite his careful wording and indefatigable reiteration, there was no accord over how to read the term "gender." This aggrieved Money; as one commentator put it, he was "irritated by the failure of most of us to stick to his original precise meanings."[53] Increasingly from the 1970s onward—and partly in reaction to second-wave feminism, as we have noted in chapters 1 and 6—Money objected that gender role and identity were used "without unanimity as to their definition," with the result that "not everyone understands them the same way."[54] In the closed linguistic world of his work, Money's attempts to achieve unanimity did not encompass the reappraisal of his definitions in view of arguments made by others. Even when, after Stoller, Money added a definition of gender identity alongside that of gender role, he did so to emphasize that these were really the same phenomenon, complaining that "people proved incapable of conceptualizing their essential unity."[55] Money's concern about this state of affairs was not only that other people were wrong, but that their errors might acquire the very scientific status that he sought. He protested that "anyone who wants to become his own expert and furnish himself with his own definitions [of gender identity and its disorders] may do so."[56] This was an extraordinary complaint from a writer whose primary strategy for the assertion of authority was the creation of neologisms.

## A WORLD OF LANGUAGE

Seeking to differentiate his work from that of other aspiring experts, Money turned again in the early 1980s to the practices of reading and writing, and once more with equivocal results. He made a special book. As Money later recalled, "In 1982 I made a collation of journal articles from my own bibliography and had three sets of them bound and titled *Principia Theoretica*."[57] This faux-Newtonian title is fascinating for two reasons. First, its self-aggrandizing scientificity perfectly captures Money's ambition to be seen simultaneously as one scientist among many and as a sexologist like no other. The second reason, which runs counter to the first, is that the incongruously Latinate wording of the title invokes an aspect of language with which Money struggled the most: bilingualism. We want to end by suggesting that Money's difficulty in accounting for bilingualism exemplifies the implications of his closed linguistic world.

Bilingualism posed a problem to Money, because it seemed to represent equivocation in spoken language. During the 1950s and 1960s, Money had presented bilingualism as akin to improper gender development, stating that "it is possible for an individual to establish an ambiguous gender role, just as it is possible for him to become bilingual."[58] In this regard, the critic David Rubin has observed, "Money's ultimate goal was to eradicate ambiguity in the name of promoting monolingualism."[59] From the 1970s, though, Money presented bilingualism as akin to *proper* gender development, claiming that just as "the bilingual child encounters two sets of language stimuli requiring two sets of responses, so the ordinary child receives and responds to two sets of gender stimuli," and thereby imitates either male or female behaviors.[60] In this revised explanation, the lesson of bilingualism was not unwanted ambiguity, but the establishment of what Money approvingly called "duality."[61] Even so, Money was not quite able to fit bilingualism into his theory of gender and language. The analogy between bilingualism and the imitation of a single set of "gender stimuli" relied upon a narrow definition of bilingualism to mean speaking in one language and listening in another, for instance, in immigrant children who "know how to listen to the language of the old people, but they won't use it."[62] On one occasion, Money admitted that "there are many bilingual children who listen and talk in both languages," but explained weakly that he was "picking the other example because of the analogy to be found with gender identity."[63] We might call this a monolingual use of bilingualism, not just for its disregard of individuals who actually speak two languages, but also for its attempted reduction of the word "bilingualism" to a single, unequivocal phenomenon.

Bilingualism challenges the traditional project of cartography, for in Rubin's words, it "reveals the promise of border crossing" and destabilizes ethnocentrism.[64] In the context of Money's work, it raises the question of how different his account of gender might have been if he had taken as a model those "bilingual children who listen and talk in both languages." Moreover, beyond the private garden of neologisms that Money doggedly constructed, the world is more than bilingual; it is multilingual—in academic disciplines, feminisms, genders, and sexualities, as well as in national languages. Its power relations are multilingual and multifarious too: we are not proposing a simplistic celebration of plurality. But we think that it is still possible and important to critique the lonely singularity of Money's œuvre. To some readers, his repetitive publications and circular claims may seem merely arrogant, indicative of the "somewhat prickly personality" recounted by his colleagues.[65] On the basis of the readings that we have presented in this book, we suggest instead that Money's work articulates a lifelong sadness over the difficulty of dwelling in a linguistic world with others, and the frustration that followed from an assumption that only with a common language can one share a world.

## NOTES

1. John Money, *Sin, Science, and the Sex Police: Essays on Sexology and Sexosophy* (Amherst, NY: Prometheus, 1998), 283.

2. John Money, *Gendermaps: Social Constructionism, Feminism, and Sexosophical History* (New York: Continuum, 1995), 12.

3. Money, *Sin, Science, and the Sex Police*, 283.

4. Money, *Sin, Science, and the Sex Police*, 101, 166, 167.

5. "Biological Imperatives," *Time*, January 8, 1973, 34. These pictures first appeared in John Money, Joan G. Hampson, and John L. Hampson, "Hermaphroditism: Recommendations Concerning Assignment of Sex, Change of Sex, and Psychologic Management," *Bulletin of the Johns Hopkins Hospital* 97 (1955): 284–300, 292, 293.

6. John Money and Patricia Tucker, *Sexual Signatures: On Being a Man or a Woman* (Boston: Little, Brown, 1975), 5.

7. Beryl Markham, *West with the Night* [1942] (Eastford, CT: Martino Fine, 2010), 245.

8. John Money, *Sex Errors of the Body: Dilemmas, Education, Counseling* (Baltimore: Johns Hopkins Press, 1968), 46. Although Money claimed in this passage that such reassurance arose from "participating in and understanding the meaning of a decision," the power dynamics in consultations about treatment were unlikely to have been equitable; and Money suggested that a person who had received medical treatment in infancy would, when older, need "some explanation about his or her condition, if only to rationalize the need for medical checkups"—not because, for example, they had a right to know. See John Money, "Psychologic Evaluation of the Child with Intersex Problems," *Pediatrics* 36 (1965): 51–55, 52.

9. John Money, Reynolds Potter, and Clarice S. Stoll, "Sex Reannouncement in Hereditary Sex Deformity: Psychology and Sociology of Habilitation," *Social Science and Medicine* 3 (1969): 207–16, 211.

10. J. B. Harley, "Deconstructing the Map," *Cartographica* 26, no. 2 (1989): 1–20, 4.

11. John Money and Anke A. Ehrhardt, *Man and Woman, Boy and Girl: The Differentiation and Dimorphism of Gender Identity from Conception to Maturity* (Baltimore: Johns Hopkins University Press, 1972), 261.

12. John Money and Mark Schwartz, "Biosocial Determinants of Gender Identity Differentiation and Development," in *Biological Determinants of Sexual Behaviour*, ed. John B. Hutchison (Chichester: John Wiley, 1978), 765–84, 782.

13. Rosalyn Diprose and Robyn Ferrell, introduction, in *Cartographies: Poststructuralism and the Mapping of Bodies and Spaces* (Sydney: Allen and Unwin, 1991), viii–xi, ix.

14. Judith Butler, "Melancholy Gender/Refused Identification," in *The Psychic Life of Power: Theories in Subjection* (Stanford, CA: Stanford University Press, 1997), 132–50.

15. John Pickles, *A History of Spaces: Cartographic Reason, Mapping, and the Geo-Coded World* (London: Routledge, 2004), 89.

16. Money and Ehrhardt, *Man and Woman*, xii, xi.

17. John Money, "Determinants of Human Gender Identity/Role," in *Handbook of Sexology*, ed. John Money and Herman Musaph (Amsterdam: Elsevier/North Holland Biomedical, 1977), 57–79, 63. Critiques of the anachronistic standards that Money used to assess gender include Rebecca M. Jordan-Young, *Brain Storm: The Flaws in the Science of Sex Differences* (Cambridge, MA: Harvard University Press, 2010), 118.

18. John Money, "Sin, Sickness, or Status? Homosexual Gender Identity and Psychoneuroendocrinology," *American Psychologist* 42 (1987): 384–99, 385.

19. By the time of Thanksgiving in the year that Money's article was published, 25,644 Americans were known to have died of AIDS-related diseases. See Douglas Crimp, "AIDS: Cultural Analysis/Cultural Activism," in *AIDS: Cultural Analysis, Cultural Activism*, ed. Douglas Crimp (Cambridge, MA: MIT Press, 1988), 3–16, 11.

20. John Pickles, "On the Social Lives of Maps and the Politics of Diagrams: A Story of Power, Seduction, and Disappearance," *Area* 38 (2006): 355–64, 349.

21. Pickles, *A History of Spaces*, 89.

22. Money, Potter, and Stoll, "Sex Reannouncement," 214.

23. Emily Grabham has critiqued the use of similar imagery in more recent medical depictions of sexual development. See "Bodily Integrity and the Surgical Management of Intersex," *Body and Society* 18, no. 2 (2012): 1–26, 18–12.

24. As Anne Fausto-Sterling has commented, "To inform a three-year-old girl about her prospective clitoridectomy Money and his co-workers tell her that 'The doctors will make her look like all the other girls.' If the surgery results in genitalia that looks like those shown in Money and Ehrhardt's book [*Man and Woman, Boy and Girl*], then these particular psychologists are in need of an anatomy lesson!" (*Myths of Gender: Biological Theories about Women and Men*, rev. ed. [New York: Basic, 1992], 138).

25. John Money, "Human Behavior Cytogenetics: Review of Psychopathology in Three Syndromes—47,XXY; 47,XYY; and 45,X," *Journal of Sex Research* 11 (1975): 181–200, 196.

26. Marcus A. Doel, "The Obscenity of Mapping," review of *A History of Spaces: Cartographic Reason, Mapping, and the Geo-Coded World*, by John Pickles, *Area* 38 (2006): 344–45, 344.

27. Doel, "Obscenity of Mapping," 344.

28. Doel, "Obscenity of Mapping," 344.

29. Harley, "Deconstructing the Map," 15.

30. John Money, *Gay, Straight, and In-Between: The Sexology of Erotic Orientation* (New York: Oxford University Press, 1988), 118.

31. Money and Tucker, *Sexual Signatures*, 3.

32. J. Hillis Miller, *Topographies* (Stanford, CA: Stanford University Press, 1995), 7.

33. Miller, *Topographies*, 7.

34. Kenneth Zucker, quoted in Benedict Carey, "John William Money, 84, Sexual Identity Researcher, Dies," *New York Times*, July 11, 2006, section B, 8.

35. Richard Green, "John Money, Ph.D. (July 8, 1921–July 7, 2006): A Personal Obituary," *Archives of Sexual Behavior* 35 (2006): 629–32, 630.

36. Vern L. Bullough, "The Contributions of John Money: A Personal View," *Journal of Sex Research* 40 (2003): 230–36, 235.

37. Leonore Tiefer, review of *Love and Love Sickness: The Science of Sex, Gender Difference, and Pair-Bonding*, by John Money, *Signs* 7 (1982): 914–17, 917.

38. Daniel C. Tsang, "Policing 'Perversions': Depo-Provera and John Money's New Sexual Order," *Journal of Homosexuality* 28 (1995): 397–426, 412–13.

39. Lesley Rogers, "The Ideology of Medicine," in Dialectics of Biology Group, *Against Biological Determinism*, ed. Steven Rose (London: Allison and Busby, 1982), 79–93, at 86.

40. Anke A. Ehrhardt, "John Money, Ph.D.," *Journal of Sex Research* 44 (2007): 223–24, 223.

41. John Money, *The Psychologic Study of Man* (Springfield, IL: Charles C. Thomas, 1957), 3.

42. Money, *Sin, Science, and the Sex Police*, 173.

43. John Money, review of *The Intersexual Disorders*, by Christopher J. Dewhurst and Ronald R. Gordon, *Journal of Nervous and Mental Disease* 152 (1971): 216–18, 216.

44. Compare the definitions in Money and Ehrhardt, *Man and Woman*, 4, with those in John Money, "Hermaphroditism, Gender, and Precocity in Hyperadrenocorticism: Psychologic Findings," *Bulletin of the Johns Hopkins Hospital* 96 (1955): 253–64, 254.

45. Money and Tucker, *Sexual Signatures*, 9.

46. Brenda Love, *Encyclopedia of Unusual Sexual Practices* (London: Abacus, 1995). Money reciprocated by offering the following endorsement for the book: "For promotions and publicity, you may quote me as follows: Whereas an *Oxford English Dictionary of Sexological Terms* has yet to be compiled, in the meantime this book is unmatched in its coverage of what is called on the street kinky sex, in the courtroom, perversion, and in the clinic, paraphilia. There is information here to benefit equally the specialist and the beginner in search of self-understanding" (letter, JM to Abacus Press, cc'ed to BL, May 26, 1992).

47. Money, *Sex Errors*, 62; Money and Tucker, *Sexual Signatures*, 112–13.

48. John Money, "Hermaphroditism," in *The Encyclopaedia of Sexual Behaviour*, vol. 1, ed. Albert Ellis and Albert Abarbanel (London: Heinemann, 1961), 472–84, 474 (hereafter "Hermaphroditism [encyclopedia entry]").

49. John Money, "Gender Role, Gender Identity, Core Gender Identity: Usage and Definition of Terms," *Journal of the American Academy of Psychoanalysis* 1 (1973): 397–402, 401.

50. Money, "Hermaphroditism," 254.

51. Robert J. Stoller, "A Contribution to the Study of Gender Identity," *International Journal of Psycho-Analysis* 45 (1964): 220–26.

52. Money and Ehrhardt, *Man and Woman*, 4.

53. John Bancroft, "John Money: Some Comments on His Early Work," in *John Money: A Tribute*, ed. Eli Coleman (Binghamton, NY: Haworth, 1991), 1–8, 6.

54. John Money, "Gender Role," 397; Money and Tucker, *Sexual Signatures*, 9.

55. John Money, "The Development of Sexuality and Eroticism in Human Kind," in *Heterotypical Behaviour in Man and Animals*, ed. M. Haug, P. F. Brain, and C. Aron (London: Chapman and Hall, 1991), 127–66, 132.

56. John Money, "Gender Role," 401.

57. John Money, *Venuses Penuses: Sexology, Sexosophy, and Exigency Theory* (Amherst, NY: Prometheus, 1986), 3.

58. John Money, "Hermaphroditism [encyclopedia entry]," 477.

59. David A. Rubin, "'That Unnamed Blank That Craved a Name': A Genealogy of Intersex as Gender," *Signs* 37 (2012): 883–908, 901.

60. Money and Ehrhardt, *Man and Woman*, 163.

61. John Money, "Paraphilias," in *Handbook of Sexology*, ed. John Money and Herman Musaph (Amsterdam: Elsevier/North Holland Biomedical, 1977), 917–28, 923.

62. John Money, "Prenatal Hormones and Postnatal Socialization in Gender Identity Differentiation," in *Nebraska Symposium on Motivation*, ed. James K. Cole and Richard Dienstbier (Lincoln: University of Nebraska Press, 1973), 221–95, 275.

63. John Money, "Prenatal Hormones," 275.

64. Rubin, "That Unnamed Blank," 899.

65. Bullough, "Contributions of John Money," 235.

# INDEX